The
SUPPLY CHAIN
Manager's Problem-Solver

Maximizing the Value of Collaboration and Technology

The SUPPLY CHAIN *Manager's Problem-Solver*

Maximizing the Value of Collaboration and Technology

Charles C. Poirier

S^{t}L
ST. LUCIE PRESS

A CRC Press Company
Boca Raton   London   New York   Washington, D.C.

Library of Congress Cataloging-in-Publication Data

Poirier, Charles C., 1936-
  The supply chain manager's problem-solver : maximizing the value of collaboration and technology / by Charles C. Poirier.
     p. cm. – (The St. Lucie Press)
  Includes bibliographical references and index.
  ISBN 1-57444-335-6
    1. Business logistics. 2. Production management. 3. Industrial procurement. 4. Materials management. I. Title.

HD38.5 .S639 2002
658.7–dc21   2002024856

This book contains information obtained from authentic and highly regarded sources. Reprinted material is quoted with permission, and sources are indicated. A wide variety of references are listed. Reasonable efforts have been made to publish reliable data and information, but the author and the publisher cannot assume responsibility for the validity of all materials or for the consequences of their use.

Neither this book nor any part may be reproduced or transmitted in any form or by any means, electronic or mechanical, including photocopying, microfilming, and recording, or by any information storage or retrieval system, without prior permission in writing from the publisher.

The consent of CRC Press LLC does not extend to copying for general distribution, for promotion, for creating new works, or for resale. Specific permission must be obtained in writing from CRC Press LLC for such copying.

Direct all inquiries to CRC Press LLC, 2000 N.W. Corporate Blvd., Boca Raton, Florida 33431.

Trademark Notice: Product or corporate names may be trademarks or registered trademarks, and are used only for identification and explanation, without intent to infringe.

Visit the CRC Press Web site at www.crcpress.com

© 2003 by Charles C. Poirier
St. Lucie Press is an imprint of CRC Press LLC

No claim to original U.S. Government works
International Standard Book Number 1-57444-335-6
Library of Congress Card Number 2002024856
Printed in the United States of America 2 3 4 5 6 7 8 9 0
Printed on acid-free paper

Dedication

For the many friends and associates who have helped guide my thinking
and assisted me with applications;
and for the continuing support of my family.

Preface

Around the world, business firms have been embracing supply chain management as the next step in process improvement and profit enhancement. As companies merge their existing improvement efforts with supply chain, they focus on the end-to-end process steps that define how products and services are created and delivered to business customers and ultimate consumers. Along the way, they find new opportunities to reduce costs, better utilize assets, and build new and profitable revenues, while delighting targeted customers. After nearly a decade of implementation, the results are becoming clear. The effort works, but there are complications encountered along the way, and large gaps have appeared between industries that were early to embrace supply chain and those still beginning to implement the inherent concepts.

The high-technology, aerospace, and automotive industries are substantially in front of forest products, construction, and industrial products and can point to significant savings at such firms as Dell, Cisco, Sun Microsystems, Boeing, Lockheed, Toyota, and Ford. In-between lie a number of industries exhibiting various levels of understanding and progress with supply chain management. All can report on some achievements, but rarely to the same degree. The differences are a function of understanding the potential value, applying the concepts properly, and resolving the complications.

Across industries, the leading business firms realize supply chain optimization, advanced supply chain management, virtual logistics, e-commerce, digital communication systems, and other modern tools of supply chain are essential factors in getting to the next level of success. Others seem unconcerned about their lack of progress as they labor under the misguided perception that focusing only on internal excellence is the key to business prosperity. This strategy only works for firms engaged in supplying basic commodities or operating in a particularly narrow focus niche business.

At the same time, many of the companies hot into the pursuit of supply chain find another dilemma. After progressing with the concepts for several years, they reach a point of diminishing returns on their efforts. Following early successes, particularly in better purchasing, lower transportation costs, more efficient order management, and improved inventory control, they discover the low fruit has been picked. It then takes a stronger effort to get at the total savings offered by a full network approach to supply chain. Now they face the cultural barriers that inhibit true collaborative effort across the network of linked participants in supply chain, and progress stalls.

Some companies are not misled by these conditions and do not assume they have reached the full potential of a focus on supply chain. They appreciate the need for and value of technology applications, designed to enhance the supply chain effort. They learn to work with willing supply chain constituents to find extra values, which have eluded their relationships. Cycle time reduction, joint product development, joint asset utilization, and virtual logistics systems become part of the effort.

Unfortunately, the recorded results of such applications contain a mixed bag of outcomes. For each notable success, a failure is recorded and a firm has its interests dampened. The recent demise of many of the dot-com entities that were once the darlings of the supply chain world exacerbates the situation by casting a shadow of doubt over any purported technology-based organization or cyber-based application. That is an unfortunate circumstance, because leveraging Internet technologies to improve a business remains a very sound concept.

Use of Internet-based technologies and digital communication across extended enterprises is here to stay and will be a part of future processing for firms leading their industries. The key is finding the right combination of technology and application, which enhances the elements of supply chain that set one network apart from another in the eyes of the customers and consumers. Business-to-business commerce, enabled by cyber techniques, needs to be a central concept in any future business strategy. Make no mistake — supply chains of the future will be technology chains, linking companies from each end of the chain in a digital framework that makes the most sense to the intended buyers.

Companies on the leading edge of the effort are actively trying to find the answers and solutions to the impediments, as they seek the optimum solutions first, and then the right data applications and software that will improve their business processes. They are in pursuit of determining how to link those applications with partners in what becomes a value chain network. These pathfinders, however, face barriers to solid progress, as members of their own organizations stall efforts due to poor understanding of the correct sequence of activities and the crucial need for a compelling design to guide the effort.

Eliminating the mistakes, dealing with the pitfalls, battling the barriers, and braving the criticism cast by the non-believers include art and science. It becomes the means of leading the pack or forever being a follower. In this book, we will take a hard and experienced look at the art of developing meaningful supply chain collaboration and the science of using technology to enable the resulting systems. Using first-hand knowledge and not hearsay, we will explain the means to make an effort succeed. We will consider the 12 biggest mistakes that exist for each of the company categories mentioned. We will provide generic examples and specific instances from many industries and companies. More importantly, we will consider solutions, complete with actual case studies and action stories showing how collaboration and proper application of Internet-based technologies are typically the keys to success. Along the way we will consider what it takes to succeed.

The work will reflect years of in-depth research, interviews with hundreds of key players across the supply chain spectrum, case study analysis, and actual hands-on experience spanning a wide number of industries and firms. The book is rich in detailing the 12 mistakes, how they are made, and how they can be overcome. In the process, we will define a course of action that ensures better performance. Business examples will be documented for each chapter. The reader will be given a framework for understanding the mistakes and how to use the solutions to build a more effective supply chain improvement effort, or to enhance an existing effort that needs to be recharged to get to the next level of progress.

This book will be of interest to anyone engaged in, or considering being engaged in, supply chain management, particularly those supply chain professionals seeking the answers to their many problems of implementation. It will have meaning for executives looking for how to get to the next level of performance improvement and managers charged with getting the intended solutions and returns on effort. It should be of interest to suppliers and users of software as we see how to get the right sequence of implementation and raise the probability of success. Academicians will find value here as a guide for how to overcome the many obstacles that stand between good intentions and superior achievements. There should also be global interest as supply chain continues to circle our world and bring new efficiencies to those willing to make the special effort to build a dominant value chain network.

Supply chain management is a hot topic, and much remains to be accomplished. There is still time to gain the high ground in an industry, as no firm has mastered all of the techniques and tools to claim the dominant position. There are leaders and followers, but opportunities exist to take the inherent concepts to a position of leadership. Overcoming the 12 most common mistakes can only help in that quest.

Acknowledgments

Recognition must be given to those who helped in bringing this book to fruition. There are more names than can be listed, but special thanks go to Alex Black, Michael Bauer, Chet Chetzron, Travis Cole, Erika Dery, Cheryl Doggett, Jean Eske, Roger Doty, Drew Gant, Steve Georghakis, David Groener, Deb Hageman, Ken Hill, Pat Holmes, Michael Holzer, Bill Houser, Dave Howells, Marty Jacobsen, Gary Jones, Steve Keener, Larry Maloney, Terry Newsome, Frank Quinn, Steve Reiter, Brad Scheller, Elsbeth Shepherd, Christopher Slee, Bob Trauner, Chuck Troyer, Ian Walker, Chuck Weiss, and Alan Weyl.

About the Author

Charles C. Poirier is a partner with the Supply Chain Solutions practice of Computer Sciences Corporations (CSC), one of the world's largest information technology and management consulting firms. He is a regular contributor to domestic and international conferences and seminars on subjects ranging from supply chain optimization and electronic commerce to finding hidden values throughout business enterprises and associated partnering opportunities. He has published articles and papers in many of the leading business journals, dedicated to supply chain process improvement, collaboration across inter-enterprise networks, and business enhancement through the application of enabling technology.

Mr. Poirier has held a variety of management positions, including senior vice president of manufacturing and marketing at Packaging Corporation of America. He was also group manager of a major business unit of that company. His background includes direct management experience in productivity, quality, cost containment, business strategy, mergers and acquisitions, training, sales and marketing, and information technology. His previous publications include *Business Partnering for Continuous Improvement* (Berrett-Koehler), *Avoiding the Pitfalls of Total Quality Management* (ASQC Press), *Supply Chain Optimization* (Berrett-Koehler), *Advanced Supply Chain Management* (Berrett-Koehler), *E-Supply Chain* (Berrett-Koehler), and *e-Business: The Strategic Impact on Supply Chain and Logistics* (Council of Logistics Management).

Contents

1. Introduction: An End-to-End Value Chain with Cyber Enhancements 1
2. Mistake 1: Lack of Leadership Vision 11
3. Mistake 2: Using the Wrong Metrics 27
4. Mistake 3: Aversion to External Advice 39
5. Mistake 4: Focusing Only on the Bottom Line 55
6. Mistake 5: Poor Customer Relationship Management 65
7. Mistake 6: Not Focusing on the Consumer 79
8. Mistake 7: Misunderstanding the Internet 93
9. Mistake 8: Lack of Collaboration Across the Extended Enterprise 109
10. Mistake 9: Weak Global Concepts 125
11. Mistake 10: Absence of Advanced Sourcing Applications 149
12. Mistake 11: Dealing Incorrectly with the Existing Culture 173
13. Mistake 12: Not Trusting the People You Need to Trust 191
14. Conclusions: The Path Forward 209

References 229

Index 233

1 Introduction: An End-to-End Value Chain with Cyber Enhancements

Before there was supply chain, there was always some form of business effort designed to cut costs to the bone. Once in a while, business people even paid attention to the quality of their output, but mostly the focus was on being the lowest cost producer in an industry, so you could bring low prices to a market and still make the most money. Every reader of this book probably has labored for a time under one or more three-letter acronym that stood for the latest continuous improvement effort. Some lasted longer than others, but the drive was the same — to wring every possible dollar out of the cost of doing business.

Then we discovered supply chain, and realized there was an "umbrella" process under which the best features of the previous continuous improvement efforts could be merged with a focus on end-to-end processing that resulted in customer satisfaction. It was a time when we saw an opportunity to redesign and reengineer the linked process steps, all the way from initial raw materials to delivery of finished goods and services. While pursuing this opportunity, the idea of supply chain optimization developed — the chance to bring all process steps to a best practice level and thereby optimize the total effort. Emphasis on cost reduction would not go away under this effort, but a look would be given to each step under the umbrella to find the best way to satisfy customers, even if one or more steps incurred higher costs. The first problem was to determine what end-to-end meant for a particular firm. The second was to determine the scope of the ensuing improvement effort.

For the first matter, the preference is to use as broad a definition as possible, so most process steps are included and, therefore, the greatest opportunity for improvement is considered. In that sense, supply chain begins at the front end (upstream) with earthbound products such as ores, minerals, grains, chemicals, and so forth. The chain continues through delivery to a converting or manufacturing operation, and on to warehouses, distribution centers, stores, and final consumption by a consumer.

In the modern sense, the chain is complete at the back end (downstream) if the consumer is satisfied with the delivery. If not, and something is returned, the chain continues in a reverse manner. With returns from Internet buying, for example, running as high as 30%, reverse logistics is very much a part of some supply chains. To optimize the effort, a company must define the breadth of its end-to-end processing so the people chasing improvement know where to begin and where to stop. Then the intermediate steps can by mapped and effort brought to the appropriate points for improvement.

For the second problem, the firm must define the scope of activities that will be considered. The preference here is for a scope that begins locally and expands in both directions across the full breadth of the end-to-end processing. That means teams begin working on internal functions and processes to clean up existing problems, mistakes, and errors, so an internal efficiency is established. The internal house must be in order before venturing into the external environment. Then the focus moves outside to look at how every hand-off in the processing can be improved and made as effective as possible, for the needs of the customer. Eventually, the effort spans a full network — an extended enterprise — and applies the appropriate cyber-based technologies to establish the most effective value chain in the eyes of the desired customers and consumers.

With Supply Chain Comes an Evolutionary Procedure

With a proper definition of the scope and breadth of supply chain, a firm is in position to determine which constituents can influence the optimization opportunities and begin working on the end-to-end improvement effort. Every firm goes through an evolution in this process, beginning inside the organization. This internal view is important and must be accomplished, particularly to bring together the disparate parts of a typical business so an efficient supply chain system can be established within the walls of the organization. Once that level of progress has been achieved, the firm can look externally for the partners with which the higher levels of advanced supply chain management can be achieved. The journey continues for several more stages, as the linked firms now look for the means to enhance their relationship and optimize efforts with technology and collaboration.

To summarize the evolution, progress moves through the following levels:

- Level 0 – Pre-Supply Chain – Continuous Improvement – Bring Forward the Good Practices

    Every firm has some form of improvement effort under way before embarking on supply chain. It is nearly impossible to cancel such efforts, and doing so would be a mistake. The improvements made must be a part of the new focus. The caveat is to be wary of putting so much emphasis on supply chain that the firm risks throwing away the good practices, systems, and methodologies developed under the Level 0 or starting position when embarking on what is viewed as a new and comprehensive effort. The temptation will be to start with a fresh approach to process improvement. That is a mistake. Bring the best of what came from the continuous improvement effort forward is the correct thinking. And merge it with a focus on the end-to-end processing that takes place as the firm goes to market.

- Level 1 – Internal/Functional – Focus on Sourcing and Logistics; Consider Business Unit Interaction and Potential Organizational Synergies

    In the first level of supply chain evolution, the firm invariably works on an internal basis, still seeking cost improvement while calling it supply chain effort. The focus is on functional improvement, usually within a specific business unit, as departmental silos and independent operating units dot the landscape. Little cross-organizational cooperation exists in this early level, but real savings are possible, particularly from sourcing and logistics. The sourcing supply base is reduced, volumes are leveraged, and costs decline. So long as quality is not impaired, savings can be significant and funding is created to continue the effort into the other levels of progress.

    In the area of logistics, transportation costs are reduced, warehouse space is matched with need, and inventories are put under scrutiny for possible reduction. Most firms finish Level 1 with an improved order entry and order management system that eliminates the errors that confound such systems, and speeds the cycle time from receipt of order to receipt of cash payment. Some firms begin an effort to apply these improvements across business units to establish a basis for intra-enterprise interaction and cooperation.

- Level 2 – Internal/Cross-Functional – Focus on Internal Excellence – Break Down the Internal Walls and Begin Corporate Integration

   In the second internal level of progress, the walls that typically separate parts of the organization and inhibit leveraging the full power of the scale of operations are broken down. Now the company starts cooperating with itself and cross-functional effort appears. Separate business units come together to see what they buy in common, what they process and ship, and to determine where the opportunities exist to collaborate, without harming market capabilities or functional excellence.

   Some form of shared services based on aggregated demand is created within the firm, to take advantage of the full size of the organization. Transportation needs across the firm are studied to see where synergies can be applied. Sharing of best practices starts to move across functions and units. The seeds of advanced technology are planted here, as a communication intranet is established and software introduced to enhance planning and scheduling. It is a time of polishing the sword for the ensuing battle — getting the organization primed with respect to supply chain process steps that are vital to market success.

- Level 3 – External/Network Formation – Focus on the Customer through Collaboration with Selected Partners

   Although every firm will claim it has a strong focus on the customer, that does not happen until a move is made to an external viewpoint. A strong cultural wall stands between Levels 2 and 3; a wall that schools all effort should be focused on internal or corporate excellence. Customers are only a by-product of the effort. With an external view, the firm can move forward with the help of partner collaboration.

   Now the company seeks out the willing constituents of the supply chain that can assist in finding the next level of improvement. An extended enterprise perspective is brought to the discussions, as the firm realizes it is only a part of the network of companies focused on a particular customer or consumer group. Together these allies focus on customer satisfaction and bring alignment to supply chain efforts, so a distinctive advantage is gained in the eyes of those customers.

- Level 4 – External/Value Chain – Focus on the Consumer with Partners and Establish Inter-Enterprise Synchronization

  As collaboration succeeds and technology is used as a key improvement tool, the linked firms move into an industry leadership position, where a value chain constellation begins to form. This entity is a set of firms cooperating as an extended enterprise to dominate a particular market or industry by virtue of having the delivery system of choice in the eyes of the desired business customer.

  But a new dimension is added. Realizing that any supply chain ends with consumption, the focus moves to a targeted end consumer group. Now network resources move from attention to the bottom line (cost reduction) to the top line (new revenues in the desired market sectors). The supply chain becomes a value chain effort in this level, as enough information is shared to pinpoint all the costs and values from end-to-end of the network and partners focus on how they can optimize all the process steps. Working together, members of the value chain begin to synchronize efforts across the inter-enterprise network. That means the firms establish alignment of the supply chain process steps into a single, logical, extended enterprise, operating as a fully linked and optimized end-to-end network from suppliers to consumers.

- Level 5 – Full Network Connectivity – Focus on Cyber Technology as the Value Chain Enabler and Achieve Network Optimization

  The final level of progress is more theoretical than factual because of the limited number of firms that occupy this space. An area where full network connectivity has been achieved, in which all of the important transactions are visible online. Partners are sharing vital information electronically and bringing unprecedented low cycle times to the processing that takes place across the full network. Inventories are viewed on a real-time basis, forecasting error has been reduced to workable levels or banished in favor of direct linkage to consumption, transportation is a virtual effort taking advantage of all the modes in a system, and new products come out in a fraction of typical time frames with a higher possibility of success. The opportunity to make savings while generating new revenue is possible for all parties in the value chain.

The Progression Is Not Easy

Exhibit 1.1 illustrates the evolution we are describing. Once a firm determines to merge its existing improvement effort with a drive for supply chain optimization, the first level of the evolution is focused on integrating best processing within the enterprise. This effort requires bringing an understanding to the firm that the separate units and functions are better served by working out the best practices and applications using the full leverage of the organization (not an easy task).

The second level brings focus to corporate excellence. The cultural wall between levels has to be surmounted to get to the next level of partner collaboration. Most firms are working in this area, trying to achieve a satisfactory Level 2 improvement. Some companies will find they have business units with footprints in multiple levels, as a few units will move faster than others.

Using modern, cyber-based technologies and collaborative commerce, linked firms using partner cooperation can move along to the advanced levels of the progression. Many innovators, or pathfinders, are forging new levels

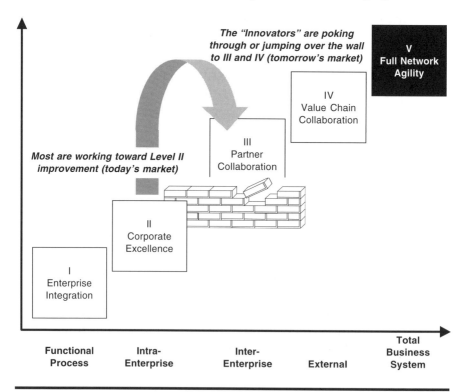

Exhibit 1.1 The supply train evolution.

of improvement as they poke through or jump over the inhibiting cultural wall. These firms work with other partners in their value chain to find the means to continue to improve performance, while discovering the means to differentiate the network in the eyes of the desired customers. With the total linking of the network electronically, an unprecedented level of full network agility can be brought to what becomes the value chain constellation, and markets can be dominated.

This evolution is the essence of supply chain progression as it has been played out in industry after industry and company after company. It is the framework we will use to analyze what goes right, what gets in the way, what goes wrong, and how it can all be sorted out and resolved for a successful effort.

Results Can Be Significantly Better for the Effort

The result for those firms making the most progress is an optimized end-to-end value chain with the best technology enhancements, well defined in scope and breadth, with eager partners working together to achieve and sustain a position of market dominance. The differences between those in the lead and those following can be dramatic. Research from Computer Sciences Corporation (CSC), based on responses to an annual survey of organizations engaged in supply chain, indicates a significant difference in results achieved between those firms in good supply chains and those in poor supply chains (*Source:* Computer Sciences Corporation, El Segundo, CA).

How To Harvest the Fruit Becomes the Issue

As businesses began to chase supply chain, to achieve the kind of benefits illustrated in Exhibit 1.2, a realization had to be established. The organization itself can be the biggest inhibitor to progress. Most firms, of any scale, simply do not want to share amongst internal units to find optimized paths to market. They strongly prefer to adopt or pursue what they believe are solid industry- or market-based practices, time honored in acceptance, even when results indicated suboptimum progress was being achieved. Only with forceful leadership are most of these firms able to move forward.

With progress came the understanding that a desire to accept external advice was also anathema to practitioners. As this inhibitor was overcome and real progress made with the help of willing partners, firms reached another realization — that only a handful of people truly understood what the technology implications and full cyber-based opportunities were

Exhibit 1.2 Effectiveness of Overall Supply Chain Effort Does Make a Difference

Benefits	Companies in Good Supply Chain Experiences (%)	Companies in Poor Supply Chain Experiences (%)	All Companies Surveyed (%)
Increased sales	41	14	26
Cost savings	62	22	40
Increased market share	32	12	20
Inventory reductions	51	18	35
Improved quality	60	28	39
Accelerated delivery times	54	27	40
Improved logistics management	43	15	27
Improved customer service	66	22	44

Source: Computer Sciences Corporation, El Segundo, CA.

all about. These and other complications stood in the way of the kind of swift progress made by firms without these obstacles. One of the reasons the high-technology sector moved along faster than others was because it was not encumbered with so many past practices that could not be broken or changed.

Before proceeding to the specific mistakes and solutions that were encountered, it is important that we set the stage for our analysis. First, supply chain optimization is a concept that can only be approached. So much progress has been made and so many new innovations introduced that it becomes a moving target, and only a few firms come close to best practice across an entire extended enterprise. Nevertheless, the effort is extremely valuable as the magnitude of improvement to performance is potentially the difference between survival and failure.

Second, each level of progression takes time to accomplish. The necessary understanding and requisite behavioral change is not easy and requires a lot of patience and sustained executive endorsement. For some firms and industries, achievement of the most advanced levels could require a decade of effort.

Third, the rewards for the effort can truly be the difference between leading an industry in performance or perpetually following. Research underpinning this text shows those companies (leaders) implementing Levels 3 and 4 are experiencing significant advantage over other firms in their industry and are benefiting from profitable revenue growth that would not otherwise have occurred.

Fourth, companies in earlier stages (followers) will continue to see margin erosion while trying to catch the leaders. The gap between leaders and followers shows a 1- to 2-year advantage. At the same time, no firm has reached the fifth level of progress, although many are in pursuit of this position. Therefore, the opportunity exists to solve the mysteries and mistakes confronting supply chain progress, and thereby close the gap and attain dominance.

As we go forward with the mistakes and the solutions, we will outline a course of action that can move a firm to that position. What is required is a solid understanding of what "end-to-end" means for a particular organization, what the scope of improvement effort encompasses, and how to merge technology with sound business practices to become the chosen network of supply by a particular business customer or end consumer group.

Summary

Supply chain is the new business performance improvement effort of choice, and for good reason as the results are proving. It is the umbrella process under which a firm can merge the best of its previous continuous improvement efforts to gain the next level of better performance and customer satisfaction. Unfortunately, the results are very mixed and leaders that have found the secrets to success are creating a challenging gap for the followers to overcome.

Taking best advantage of supply chain management requires defining the end-to-end process steps involved, the scope of what must be achieved, and then progressing through evolutionary levels, which will lead to market dominance. Moving through each of the intermediate levels is important to bring the organization to optimum conditions. As firms move to the external environment required for the higher level of progression, they find getting to the most advanced levels requires collaboration with a select group of supply chain partners working as a network focused on specific customer or consumer groups.

As this collaboration proceeds, the network of partners will get to the highest level with infusions of the appropriate technology and Internet-based applications. We will expand on these concepts as we now begin to pursue the many obstacles that can stand in the way of a well-intentioned effort, and explain how solutions to those obstacles can result in a firm and its partners becoming an end-to-end value chain with cyber enhancements, dominating a chosen market or industry.

2 Mistake 1: Lack of Leadership Vision

On the way to finding the secrets to a successful supply chain effort, I have asked a number of senior managers, particularly at the CEO, CFO, COO, CIO, and business unit leader level, a few questions regarding how the firm got involved in supply chain. These informal dialogues have been extremely useful in helping guide firms to better results.

When the query was specifically: what caused you to adopt a focus on supply chain management, the most prevalent response was because a major customer indicated we should be working in that area. This reply indicates the urge was not internally generated, but foisted upon the company as an implied threat that impinged on future business. The customer was typically looking for benefits that would come to the customer, and there was probably not going to be much value left for the manufacturer. When the major customer seemed to be satisfied, these firms tended to lose interest in the effort.

A few leaders indicated the decision was made because they saw supply chain as a vehicle for moving at least a portion of the organization toward an advanced level of performance that would have meaning to the firm and its customers. An even smaller group stated they had studied the subject and became convinced supply chain should become an inherent part of the way the firm conducted its business and nothing short of optimization should be the ultimate objective. These latter groups seemed to make greater progress than the former. Early improvements were typically significant, the positive results could be traced to the profit and loss (P&L) statement, and the effort was continued into advanced levels. Firms in this area tend to lead others in their industry.

In-between, there were many varied responses, and the difference in the answers spelled the difference between degrees of success. Understanding the importance of this variation, the need for active and continued leadership, and the crucial role of a guiding vision throughout the effort will be the subject of this chapter. This will not be a typical call for leadership support, a well-overworked thesis. It will be a discussion of why supply chain should be a central part of business strategy, understood and supported by the CEO and most members of the senior management staff. It will also illustrate how those guiding the course of the company effort must author the vision, be certain to align with the central tenets of that vision, and remain active in pursuing all five stages of the ensuing effort, if they expect to succeed.

The Purpose Behind the Effort Correlates with Progress

To appreciate how this mistake interferes with supply chain success, we must understand that stated purposes by senior executives and business leaders will directly impact the subsequent progress made with the supply chain effort. When the questioning was pursued with my informal sample, to get a deeper understanding of the genesis behind the supply chain effort, other frequent responses included:

- *We heard about supply chain from a respected source and decided it was a logical extension of our current improvement effort.* In this case, the leadership vision was limited and responsibility quickly passed to a lower level manager or director for implementation. Progress in Levels 1 and 2 could be expected, but limited participation in higher levels would result. The purpose behind the effort was simply not clear enough to promote strong and continued progress.
- *We heard one or more of our competitors were into supply chain and making progress.* That is a very weak response and generally characterized by poor results. Too many of the traditional measures of success were used by nonsupporters to destroy the emphasis on the effort, and people inside the firm quickly assumed they were at a higher stage than the competitor. "Besides," they would tell themselves, "the competitor isn't that good anyway."
- *We read a compelling argument in an article, white paper, journal, or magazine that sparked interest.* If the interest was followed by a serious study of the subject, its implications, and how supply chain could

become a central part of the way business was conducted, chances increased for a successful effort. As an enlarged group of senior executives was brought into the supporting group, chances increased even further. This is an area where most successful action got started. A leader heard something that sounded good and became energized about the subject. If this leader also became directly involved in establishing the charter for the effort, set up a clearly defined infrastructure, and initially put the matter in the hands of a competent subordinate, chances for success increased dramatically. The purpose was clearly and openly articulated and progress followed the expected results.

- *We responded positively to a proposal put forth by subordinates or an internal committee established to study the costs and benefits.* Some firms got started because subordinates became energized by the subject and took the time to put a case before management. This was probably the second most popular venue for starting a serious effort. Many firms met with early success here, but an equal number stalled because the senior executive endorsees did not have the necessary grounding or interest in the subject to sustain direct involvement. As their time and interest went back to normal duties and priorities, the group that initiated the effort was generally left to flounder.

- *We studied the features of supply chain and viewed it as a means to making more profit, primarily by cutting costs in areas currently not under focus.* Further, we believed it would enhance our opportunities in our marketplace in the eyes of our customers, and allow us to build new and profitable revenues. This is the strongest of the reasoning found in leading firms. The purposes are basic to the firm's future, the tenets are easily expressed, and progress can be tracked. The company leaders may not have seen the need in the beginning for strong external support (virtually no firms realized this requirement at first), but they at least were prepared to re-orient the organization to something powerful in scope. The primary purposes were internally focused but leaders knew there was a larger area of unexplored opportunity that could be mined. With good results internally, these firms turned easily to the external environment for an enlarged purpose and progress in advanced levels of effort.

Unfortunately, among all of those responses, we did not hear that leaders considered supply chain as a reasonable and important central framework that would support operations and enhance strategic direction. That appreciation seemed to come later for a few of the most advanced companies, after some measure of progress had been recorded.

One conclusion from this ad hoc research was that most senior executives do not fully appreciate or understand the underlying concepts, or the direct and positive impact a successful effort can have on nearly all measures of business performance. As a result, they do not deliver the articulated purpose so important for a sustained effort. Another was that very few leaders see the need for a supply chain effort that includes a vision of the final conditions from a successful journey — one that includes external partnering and focuses on end consumers. When a company goes forward without a solid grounding in what is being attempted, an intended purpose, why it will be beneficial, and what the consequences can be from lacking leadership vision throughout the implementation of the effort, progress and results are seriously hampered.

The Potential Must Be Understood in the Beginning

To assure the ingredients for success in this area are present, it is vital that there is an understanding of the ultimate potential of what will become a very pervasive transformation process present throughout the senior management ranks. If the bosses know what the benefits can be, they generally have little trouble authoring the vision and bringing the necessary support to this transformation, even when it includes an external orientation involving alliances with many supply chain partners.

Recent results being recorded in a myriad of magazines, journals, newspapers, white papers, and books testify to the validity of supply chain management and its potential to reduce costs while generating new revenues, higher profits, and greater customer satisfaction. The challenge is to link the internal players to the effort, get the internal house in order, and then take a compelling vision to the external environment and secure the support of the necessary allies. In this way, the leading developments in supply chain are not missed (as one or more of the allies will bring attention to best practices), and the best processing can be made a part of the business model that will guide future efforts. The CEO has to be the leader in that orchestration. Surrogates may help the development, other executives will be expected to join the endorsement, but the vision and articulation has to come from the most senior sponsor possible.

Since no CEO is going to endorse something that does not have clear value, you need to show the savings for the firm first. Value in the eyes of customers

Mistake 1: Lack of Leadership Vision ■ 15

A Fully Functioning Collaborative Network Will Generate Significant Savings and Revenue Enhancements, Depending on the Scope of Deployment.

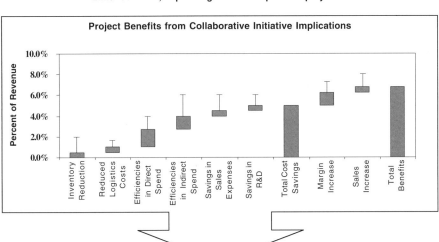

The Collaborative Network solution set looks to achieve projected cost savings on the order of 5% of revenue, with total benefits to be obtained at an estimated 7% of revenue.

Exhibit 2.1 Potential supply chain improvements.

Source: CSC projections based on benchmarked performance of comparable solution offerings.

and consumers come later, as the firm moves into the higher levels of progress. As every CEO will endorse something that adds points to the bottom line, you begin there to get the basis for endorsement. Exhibit 2.1 is an illustration often used to help orient leaders to the potential of a concerted supply chain effort. The categories and ranges of improvement opportunity are not absolute by any means, but are generally achievable through a sustained effort.

The vertical axis is intended to show the improvements that can be achieved from new profits as a percent of revenue. The categories arrayed across the horizontal axis are some of the specific areas of supply chain improvements that bring enhancement to earnings. There is no particular significance in the ordering of the categories, except that those on the left of the figure are generally achieved as the firm is moving through Levels 1 and 2 of the effort, and those to the right come as the firm moves into the advanced levels.

Beginning on the left, the potential exists to reduce **inventory** enough to bring a 0.5 to 2% improvement to earnings. The actual amount will vary depending on the starting position, but when a firm works diligently to remove inventories from the supply chain system (and not simply move them around in the network), this range is achieved. The financial improvement comes from the one-time reduction in working capital and the elimination of the carrying cost on the eliminated inventory. This area is one of the untapped sources of new profits, as most firms do not really eliminate inventory from their total supply chain. Rather, those downstream in the linkage push the inventory upstream in the misguided assumption that the cost has been eliminated. That cost will find its way into the pricing for the products and services delivered to the downstream customer. When the need for the inventory is eliminated by virtue of advanced techniques, such as online visibility of materials and goods and virtual logistics systems, the real savings appear.

As firms work first internally across business units, and then with external partners across the network to find savings in their mutual **logistics** systems, another level of savings is found. The chart shows a range of 0.5 to 1%, which could be another conservative estimate. The actual amount depends on how much effort the firm has already put into logistics. Generally, we find most firms have been successful in reducing warehouse space and outsourcing transportation, but many firms still insist on having space to hold excess inventory for emergencies and unforeseen customer needs. The majority of companies also insist on having control of the transportation and storage process because of a lack of trust in parties well suited for those purposes. Much progress remains to be gained in this area, but firms are well advised to submit a conservative expectation because of the strong pushback encountered here.

Efficiencies in **direct spend** are often touted as the largest area of supply chain opportunity. Experience schools that some care must be taken in this area. The chart shows a range of 1 to 2%. As this is an area where most of the sourcing effort is directed, low costs should already be in place. The further opportunity comes from aggregation of demand across a firm first, and then with external allies. There is also a possibility to use electronic or e-procurement techniques to reduce transaction costs and find non-traditional sources. The range shown is the result of surveys of many purchasing professionals indicating what a serious focus on reducing direct purchasing costs can provide.

Generally greater amounts of procurement savings can be found by concentrating on the **indirect spend**, where time and limited resources often interfere with getting to best practices. With the introduction of Web-based search and buy techniques and software from Ariba, Aspect, Commerce One, Harbinger, and others, significant improvement has been found in this area, and the chart reflects a range of 1 to 3% as a percent of revenues.

As the firm brings the better information and software designed to enhance sales force effectiveness to the **selling** area, we find a potential for a 0.5 to 1.5% savings. This improvement comes from the need for less direct selling expense, reduced selling time at less attractive customers, and reductions in force. The actual amount a firm can achieve depends on how much it embraces this category as a part of its supply chain effort.

The savings range for **research and development** is indicated at 0.5 to 1%. This could be another conservative estimate. A concerted supply chain effort, particularly with willing external allies, will result in a substantial reduction of the cycle time from concept to commercial success. Experience has shown that it also dramatically increases the possibility of success with new products. Measuring the bottom line effect, however, is difficult for this category so a conservative range is in order. We will discuss this area in greater detail in Chapter 8.

The total savings for what is generally an internally focused set of efforts is then accumulated to something approaching 5%. If all of the higher portions of the ranges were achieved, a much higher percent would be gained, but this is simply not the case. Years of study, including some of the most advanced supply chains, have indicated a 5% improvement as a percent of revenue is the upper limit of a 3- to 5-year supply chain effort with focus in Levels 1 and 2.

Moving to the external environment, where an emphasis is put on topline improvement with the help of network partners, another set of categories appears. First, there is the opportunity to work together to increase **margins**. The range indicated is 1.5 to 2%. In this area, the allies work to eliminate redundant assets and take advantage of core competencies. Best practice sharing leads to further reductions in operating and delivery costs, particularly through the application of virtual logistics systems. Online visibility of inventories eliminates the need for excess safety stocks and better matches supply with actual demand. Matching service with need and segmenting customers to provide electronic enhancements allow a better connection between pricing and service to those customers.

Finally, the chart shows a 1 to 2% potential for increasing revenues by working together with network partners to get **new business**. This new source of revenue generally comes by focusing, with the help of supply chain partners, on non-traditional markets and customers, often of a global nature. With greater supply chain capabilities and willing allies, the firm reaches out to new markets on a focused basis and brings in more sales. The impact is listed in the conservative range shown, because there are new selling expenses in the beginning and many trial efforts. There will be new revenues, however, and the firm should allow for some increase to profits as a result.

When the final assessment is made, we find five to eight points that can be added to the bottom line. This is the kind of improvement that excites a leadership team to endorse what may initially be a nebulous improvement process.

From the Savings Comes the Support for Vision

With an agreement on the potential benefits, a firm can begin developing the compelling vision. That requires merging what the leadership team believes can be attained with what actions and initiatives are already underway and what can be gained with help from other firms. It also requires stating the vision in terms that are consistent with the strategic objectives already outlined for the firm. This is the point where the future of the supply chain effort will lead to a high measure of internal excellence (Level 2) or the firm jumps over the cultural wall into Levels 3 through 5 for greater achievement and customer satisfaction. Every firm has a business plan that calls for increased earnings. The question is how to enhance that plan without destroying its strategic intentions and get to the advanced levels of improvement. Generally, the new vision augments the business plan by speaking to the need for external partnering to gain further benefits, to move the focus more in the direction of customers, and to build new sales with the help of supply chain partners.

The impact of a 5 to 8% enhancement has to positively affect support for advancing the supply chain effort into a network effort, and building the important concepts and actions into the business plan. Most leading firms are doing that as they build new e-business models into the framework of their strategic initiatives and business plans. The key is to translate the potential savings into a roadmap that keeps the best initiatives from existing efforts alive, while combining them with elements of an external nature and advanced supply chain efforts.

Exhibit 2.2 illustrates a technique used to combine many efforts into a cohesive plan. Every company has some overarching objective for its business improvement efforts. It can be improving earnings per share (EPS), raising economic value added (EVA), enhancing return on net assets employed (RONAE), or any other measure that has meaning to a particular organization. The chart shows that whatever the benefits development might lead to, it is a function of cost and productivity improvement (where most supply chain effort resides), better asset utilization (a newer area of focus generally

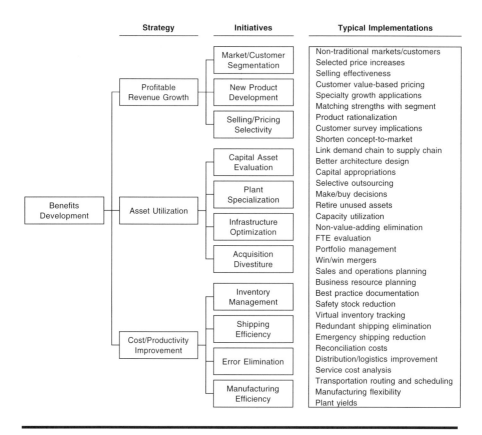

Exhibit 2.2 Path to success.

requiring network involvement), and profitable revenue growth (the collaborative focus of choice).

Each of these major categories is linked with some of the most typical areas of supply chain focus. Cost/productivity improvement, for example, is linked to inventory management, shipping efficiency, error elimination, and manufacturing efficiency in the chart. Other areas already underway can be added to that list. The final column on the right is another listing of typical implementations.

The purpose of the chart is to get supply chain constituents to sit down and take the time to list existing initiatives, add those which could come from a more collaborative effort with external partners, and those which have an electronic or cyber-based application involved. Then the list can be harmonized and related to the overall expectations of the firm, so the desired benefits developed meet the intended objectives. The results of such an effort typically lead to the elimination of lower priority and lower payback efforts that are draining valuable resources and replacing them with actions that use mutual resources provided by network partners. Experience has shown that a doubling of earning per share, for example, is possible through a concerted effort.

The Key Players Must Be Aligned with the Vision

The ideas underpinning the supply chain go beyond making improvements to sequential process steps. That is the early work for a firm under the leadership of the supply chain director. As indicated, most firms make some early progress and can show savings, usually from a focus on sourcing and logistics. To attain the advanced levels of progress, those leading the firm must decide what the company and its strategies are all about and what the enhanced future state will look like. Using the tools outlined, several key players must work with the CEO to establish what can be accomplished in terms of supply chain improvement. With that grounding, the CEO, CFO, COO, CIO, the key business unit leaders, and the head of supply chain must settle on a vision, which includes network formation and a guiding framework that will take the organization into the desired advanced level of the evolutionary process.

To create a vision that will inspire the entire organization is a tough job. It requires more than one person. To begin progress across the first two levels of the supply chain evolution, a firm can put a manger or officer in charge of the effort. It will quickly become apparent, however, that four players are crucial to ultimate success — the CEO, CFO, CIO, and the person given direct responsibility for directing supply chain progress. A fifth person is added when a visionary business unit leader is made a part of the effort, and pilots are set up to prove the concepts and show the value of the advanced stages of the evolution.

Leaving the responsibility with the supply chain leader is sufficient in Levels 1 and 2 if a dynamic, charismatic, and forceful person is selected. These qualifications are important because the early progress will depend on working with the individual business unit leaders and the heads of the functional departments to get support for what will be poorly understood concepts. Trying to pull disparate internal groups together is the first impediment to success. If it is clear the CEO has given proper endorsement to the effort, and the CFO is at least partially convinced that positive results will flow to the P&L statement, an early effort can progress.

To make progress with the intranet so vital to completing Level 2, the CIO has to become a part of the internal alliance. Without that alliance, no matter how powerful the supply chain leader, progress to the advanced levels will be inhibited. It is in Level 3 and beyond that the use of information technology and cyber-based skills are essential. The typical firm will not create excellence in this area without the direct participation of an equally skilled CIO.

With a focus on a specific range of improvement and a list of potential actions that will secure the objectives in hand, the internal group mentioned must arrange the time and facilitated help to hammer out a compelling vision that states the purposes and scope of the effort in a way that any member of the organization can understand and repeat the intentions. Trusted and valued external partners can and should be included in this discussion. The best visions are simple and compelling. They tie the future of the firm with excellence in supply chain execution and map the road to a clearly defined, enhanced future state that will lead to dominance in an industry or market. Using a specific business unit to illustrate the order of magnitude of the improvement and cost of implementation cements the necessary endorsements. Having that leader also articulate a vision for the business unit secures the cooperation necessary for moving the unit forward.

The vision must then be articulated in a way that most employees can understand what is happening and why and easily express their support. Otherwise, it is another management effort to be endured and waited out for loss of continuing support. That vision must also include a design of what the end state will be or the firm will stay forever in Level 2. That is not a bad condition if the firm supplies a commodity and participation in an external network has little meaning to company results. It is not acceptable for an organization that wants to dominate a particular market or industry.

Keeping the Focus Requires Continued Visionary Support

With so much emphasis on supply chain, it is not illogical to question why such efforts fail. Let us consider the steps in supply chain development one more time as we emphasize the need for leadership vision.

- A core group gets assigned to develop a proposition for any of the reasons mentioned in the chapter overview. Ownership of the concept and responsibility for results are usually very nebulous, but something has to happen because a senior officer said it would.
- The group gets excited by the possibilities and begins to take ownership.
- A new format or way to enhance part of the business processing is suggested and partnering within the organization is launched with specific roles and responsibilities.
- As the project scope grows, participation is expanded to accommodate more initiatives. The new players bring a different (and generally lower) level of dedication and enthusiasm.
- Among the newcomers will be a few naysayers pressed into action. Taken away from their normal duties and without a direct executive mandate, they slow or kill the progress.
  a. They were not involved in the original design so don't feel beholding to it.
  b. They do not want to change the way things are done.
  c. They do not understand the need for collaboration — suppliers are to offer services in exchange for the business, for example, is a typical concept.
  d. The system begins to slow because of the controversy. Participants look to senior executive leadership to re-endorse and solidify the imperative to proceed. In its absence, they backslide in their intensity and bicker over trivial matters.
  e. Without the leadership and intensified vision, the effort wanes and disappears. It is back to business as usual.

The lack of leadership and sustenance of the vision as critical elements in process continuance is generally lost in the leader. He or she assumes, because an early endorsement was given, that the matter is closed. It is not so for the reasons cited above. He or she gets upset at the people for wasting time/effort/hours in a failed process, but the waste of resources was due to the dulling of the intensity brought on by cultural matters not dealt with by the senior executives.

Action Study — Cisco Systems

As one of the fastest growing companies in the new business economy, Cisco Systems makes the data-networking equipment that powers the Internet. Based in Northern California, this firm has exploded from a start-up to a business worth over $100 billion in 14 years. Of Cisco's products, 95% are built-to-order, making supply chain management a necessary core competence. The firm works closely with partner suppliers to whom it outsources 55% of product fulfillment. Customers order through Cisco and the information is sent directly to suppliers who ship products directly to the customer, often bypassing much of Cisco's organization. The company's supplier relationships are so strong that 97% of the orders ship on the date promised. Cisco has reduced its cycle time on orders from 6–8 weeks to 1–3 weeks, due in part to its tight relations with suppliers.

For this firm, **instant involvement** is a keystone in the supply chain vision, and it is more than an idealized concept. It is a part of the company's basic strategy and operating philosophy. Cisco talks of its "extended enterprise" as a network of customers, suppliers, and other partners that participate in collaboration, information exchange, and relationship building. The goal is to extend business processes across corporation boundaries to build a seamlessly integrated value chain. Linkages are in place to ensure a symbiotic relationship with strategic partners. For example, information linkages ensure timely product data are given to suppliers so they can build products for Cisco. According to Bob Spiegel, a director in the Information Systems Group,

> The faster we communicate, the faster we bring product to market. For example, when an engineering change order is approved, all relevant documentation related to that change is automatically put in a ZIP file and sent to suppliers. We have a special channel on our extranet and we push this ZIP file to them. (Pearlson, 2001, p. 80)

Orders are shipped directly from these suppliers to customers. According to Spiegel, "Cisco pays for the parts and the work done by the partner, but we trust them to send the correct parts to the customers at the right time. That linkage itself means we are very closely tied to our subcontractors." Cisco has created a well-balanced, networked enterprise geared to responding rapidly to customer orders. The network, driven by instant communications over the Internet, allows suppliers to post quotes and forecasts on Cisco's Web site each quarter.

Mutual commitment is critical to Cisco's vendor relationship management (a crucial element in its supply chain effort) procedures. Not every vendor decision is made solely on price. The company has a corporate philosophy of ensuring suppliers stay profitable and, thus, in business. To implement this philosophy, the firm shares a wide variety of inventory and demand information with suppliers to help them reduce inventories and respond instantly. Cisco also pays suppliers within days, rather than weeks, which enables suppliers to reduce the costs of collecting on accounts. The company also involves partners early in the product design process to ensure products can be manufactured and that parts will be available.

Mutual commitment at Cisco means mutual investment. Managers understand that its stringent requirements for partnering can place a heavy burden on supply partners. Cisco carries its portion of the burden by investing in technology and infrastructure that benefits both Cisco and its partners. Cisco calls it the Global Networked Business Model and uses it to manage supply chain activities that integrate Cisco with its suppliers.

To make the extended enterprise a success, Cisco managers have to examine the company's capabilities before choosing supply chain partners. The selection process looks for partners with similar philosophies and complementary skills to Cisco. There are 20 to 25 strategic partners linked to the firm through electronic tools, meetings, and strategy sessions. For example, the dynamic replenishment initiative links forecasting, inventory, and backlogs to help the suppliers forecast their own production and assist in managing Cisco's overall cycle time.

The elements of this partnering are critical to Cisco's success and are an integral part of the vision that guides supply chain management for this company. That vision starts at the top of the organization. According to John Chambers, CEO, "Partnership is our heritage. Very few people in this industry partner well, so it's a huge competitive advantage." This message is driven throughout the organization and is key to a successful supply chain execution process.

Cisco's partnerships have four key characteristics:
1. Maintain the same overall vision of the industry trends and direction.
2. See short-term benefits in terms of real sales from the relationship.
3. Anticipate long-term advantages from the relationship.
4. Share similar values of being aggressive, technologically strong, and customer focused.

Instant involvement benefits Cisco and its extended enterprise in three major ways. First, Cisco senior managers are able to focus on the core competencies of product design and marketing, rather than spending time building and managing manufacturing facilities. Second, as demand increases, Cisco can ramp up production by either helping existing partners expand or by taking on new partners. This reduces the complexity of expansion to a simple task of finding appropriate partners, as opposed to investing in new plants and hiring new employees. Third, Cisco's heavy reliance on information systems integrated across the supply chain results in an extended enterprise connectivity, which appears to customers as a single, instantly responsive organization (Pearlson, 2001, pp. 80–81).

Summary

Supply chain is a hot topic that can bring five to eight new points of profit to the bottom line of a P&L statement. That is an improvement that most senior executives would exert much effort to achieve. The effort should be focused on understanding the underlying concepts and the specific potential they hold for the firm. With a firm grasp of what can be accomplished and how the effort can be merged with existing improvement efforts and the potential of using external resources, a CEO and his or her senior team must articulate a compelling vision for what will be a major transformation exercise. That vision must spell out a view of the enhanced future state and be constantly reinforced through direct involvement and rewards. If the vision is sustained throughout the effort, the biggest mistake in supply chain management will be eliminated.

3 Mistake 2: Using the Wrong Metrics

As firms drive toward optimization of their supply chains, an insidious obstacle arises. Among the many pitfalls and barriers along the way to success, one of the most difficult to overcome is using the wrong metrics. The term *metrics* in this text is interchangeable with *measurements*, to refer to the many systems and numbers applied by a company to gauge its performance, often against a preestablished standard or budget.

Since most traditional metrics quickly become oriented around the measurement of costs and profits, there should be little wonder that most supply chain efforts in a business continue to be centered on attaining internal excellence — creating the lowest cost, most profitable player in an industry. This condition will satisfy the need for a return on investment in the eyes of the stakeholders, meet the needs of the accounting group, and generally mollify those running the business. It does little to satisfy customers or help the ultimate objectives of an advanced supply chain effort that involves partnering with external constituents.

Many firms claim they understand the dilemma and some have moved to measurements more concerned with customer relations. Carly Fiorina, CEO of Hewlett-Packard, for example, has started to tie compensation for that high-technology giant to improvements in customer approval ratings. Unfortunately, the focus in the first two levels of the supply chain progression remains directly on traditional metrics, primarily cost reduction. Firms in these levels, for example, continue to pay bonuses to people engaged in supply chain improvement based on throughput and not on-time delivery and fill rates. Attention is paid to waste created in the process steps, but rarely to reducing the amount of returns from the customers. Purchasing costs are sometimes driven to levels so low that quality becomes an issue and affects manufacturing capability and product reliability. Volume and market share are rewarded rather than quality of sales or performance to commitments.

In this chapter, we will consider the dilemma of dealing with traditional measurement systems, which are not going away, while introducing a means to apply metrics related to the ultimate objectives of advanced supply chain management. We will move from improving internal conditions to meeting the external demands placed on today's business firms. Along the way, we will point out how the two camps (traditional and advanced supply chain) can be merged to create a focus on both the internal needs of the business and the external necessities of supply chain progression.

Performance Is a Function of Measurement

The importance attached to a particular metric will determine the attention of the people working for the firm and thereby influence the performance against that measure. If everyone knows that you could lose a job because you shut down a manufacturing line, then these lines will keep running regardless of the raw material building up in front of the line. Purchasing will order enough of that material to make certain there is little to no chance of down time. The space necessary to hold that material, the carrying cost, and the slow turnover of the material become ignored factors as the primary requirement is met. "No one ever lost their job because of too much inventory" is a common statement made by early level firms when we question the amount of raw material and work-in-process inventories. "But we all remember when someone lost a job because a major line was stopped because we needed something that wasn't there" is the quick second statement.

On the other hand, if you pay for customer satisfaction, you will get customer satisfaction. If you look at returns to find the reason for those returns, you will find the means to eliminate the root causes and the goods will stay with the satisfied customers. If you insist that deliveries are made within the time frame established by the customer, deliveries start to show up inside those schedules.

The problem arises when an organization brings focus to traditional measures for so long that they become a part of the culture. Any effort to move to new metrics that could have more meaning for a change effort is typically resisted, particularly if incentives and bonuses are still attached to the traditional measures. That is a condition that must be accepted as the firm moves out of Level 2 and into the advanced stages. Introductions to new metrics should be done on a patient but determined basis — usually best accomplished with a few new measures that shift the focus to the intended direction.

A typical first effort starts with a look at the direction in which analysis of flow occurs and how rewards are established. Most current measurement and bonus systems, for example, make payoffs that favor a **push** system. The company wants annual growth, so the leading metric is Net Sales. Sales personnel are compensated for the volume of orders, so they favor getting large orders, some of which contain profit and some that do not. The company also wants high profits, so the most desirable customers from a profit perspective get the most attention and best service, while others may be neglected. Manufacturing personnel are paid for getting throughput in both cases. That throughput could go directly to the best customers or, more likely, into storage. Bonuses based on units produced are prevalent in industrial and consumer products companies, with the result that these industries have traditionally high inventories as a percent of sales.

These factors drive a firm to concentrate on high-volume runs that push the goods into inventory, whether the customers really need the goods or not. Once the system is working and these goods are moving toward customers, the company has no choice but to pay for the warehousing necessary to hold the goods, to absorb the carrying cost on goods not consumed, to accept the obsolescence cost for goods never used, and to make the emergency shipments for what was really needed but not available. It is a weak, traditional system sorely in need of changing, and most supply chain efforts bring early attention to changing this focus.

The preferred focus is on a system that caters to the actual demand being created downstream in the supply chain — as close to actual end consumption as possible. With this emphasis, the supply chain becomes a **pull** system, which brings goods through the process steps in synchronization with what is being extracted. A basic premise of supply chain is that the system must match demand with supply and let consumption pull the necessary goods and services through the system. This system must bring focus to what the customer is consuming (demand), matched with available stocks (supply) in a way that the right goods are at the right place at the time of need.

In terms of metrics, the organization starts to focus on forecast accuracy, cycle times from order to delivery, out-of-stocks, late deliveries, and so forth, bringing attention to process steps related to the matching of demand and supply. Now inventories shrink closer to what is actually needed to satisfy demand, obsolescence decreases as the products most in demand are favored in planning and manufacture, warehouse space and logistics costs go down as there is less need for storage, and better end-to-end coordination occurs.

Success comes to a supply chain effort for many reasons, but it will surely occur only if coordination and best practice are achieved across the full supply chain. But this requires enormous cooperation from internal constituents. In an end-to-end effort, all parties need access to the information relating to the product, information, and cash flows. As the constituents continue to focus on their small area of responsibility, achieving optimum conditions is elusive. Manufacturing cries out for better accuracy in forecasts, sales service responds to special customer needs by interrupting schedules, purchasing pressures suppliers to match their supplies with schedule needs, and logistics managers arrange the emergency shipments to keep customers happy. It becomes a merry game of ad hoc processing.

The problem begins inside the organization where the drive for optimization, rewarded by traditional measures, can result in the deterioration of a metric important to the customer. As an example, consider what happened at one of the largest consumer goods companies in North America. This firm, with a stable full of established brands, became enamored with supply chain as a means of moving to the next level of performance improvement long before other industry members discovered the process. Unfortunately, they sustained the old measurement systems to drive performance with some unfortunate results.

The firm had traditionally rewarded manufacturing efficiency in spite of proclaiming that customer satisfaction was the more important determiner of performance. As supply chain took hold as a driving effort, attention went quickly to the area of logistics, a typical reaction for firms in Level 1. As improvements were made and costs lowered, this company decided to add a special bonus as it moved to Level 2 for achieving full truckloads before shipping case goods to customers. This seemed logical, as lowest cost conditions would occur if all trucks were fully utilized.

Since this measure drove performance, the practice caught on and trucks were held at loading docks or distribution centers until full load conditions were achieved. When the effort gained momentum and there were still trucks without full loads, the company elicited the help of other firms. They indicated to new allies that loads could be dropped off at a convenient cross-dock or the company's truck would pick up loads to fill truck capacity. Since there were obvious savings for the cooperating companies, the idea again quickly gained acceptance and the consumer goods company's full load conditions rose dramatically.

The problem occurred when these trucks were held so long waiting for the final loading that delivery dates were missed at the customer site. Eventually the problem became so bad customer complaints moved to loss of business as the customers simply changed sources, in spite of their desire to get the branded products. For the customer, meeting promises was far more important than the branded company's name and internal performance. The push system had been enhanced for the consumer goods firm, but the pull system was languishing.

The lesson is basic to getting the desired supply chain performance. Before putting a new set of metrics in place, remember that measures will drive performance; so determine what the traditional culture will do to desired performance through established practices and measures. Anticipate the needs and how important the customer is in determining good results. Then begin introducing new measures that will bring the desired new performance while expecting resistance of the move toward a pull system.

Changing Metrics Expands as an Internal Effort

Beginning with Levels 1 and 2, internal excellence in supply chain gets focused on the process steps that link sourcing, planning, production, delivery, and replenishment into a seamless process capable of meeting or exceeding customer and end consumer requirements. The typical objective is to achieve the highest possible level of basic service at the lowest total delivered cost. Traditional measures will be used to track performance and to measure how well the firm is moving toward that objective. Reduction in purchasing costs will be tracked, as will the cost of transportation, warehouse space, and inventory turns. The obstacle comes as people realize they must work together more than in the past to achieve success and the old measures do nothing to instill that kind of cooperation. In most organizations, a lack of cross-functional and cross-business unit cooperation stands in the way of getting to total efficiency.

Logistics managers, for example, when polled in a recent study conducted by Michigan State University, reported "more success in coordinating with customers than with their own purchasing and manufacturing counterparts." In the same report, purchasing managers reported "better integration with suppliers than with their own manufacturing, logistics, and marketing operations" (Stank, 2001, p. 63). In the first levels of supply chain, people pay attention to their own function or business unit and not the total effect of a cross-company supply chain. They have to be guided into accepting the importance of intra-enterprise measures.

How do you get the necessary internal coordination? Having the leadership vision discussed in the previous chapter is the first requirement. Next, there must be a strong emphasis on sharing data, usually via the newly constructed intranet that links communications between functions and business units. Sharing inventory status, sales consumption data, forecasts vs. actual consumption, and product information via this intranet provides opportunities to better manage assets and customer service. With this increased internal cooperation comes the opportunity to develop metrics focused on the customers that have meaning across the organization and the linking of those metrics to compensation.

A full exposition on appropriate measures for each function in a business is beyond the scope of this text. The action study will document some that were very useful for a consumer products firm. In general terms, a firm has to understand its own business and insist that each function determine its role in the end-to-end supply chain. Helping the supply chain effort then begins by developing metrics that aid the transition from an internal-only focus to an external view focused on customer satisfaction.

As a specific example, one function that receives constant attention as the firm moves through all levels of the supply chain evolution is procurement, alternately called purchasing and sourcing. In the early levels of supply chain, emphasis here goes to reducing the number of qualified suppliers so larger volumes can be leveraged for lower prices. Moving into the higher part of Level 2, this reduced supply base is studied to find willing and trusted sources to work on advanced partnering, bring new values, and find non-traditional ways to work together to remove mutual costs, particularly in inventory, warehousing, and transportation.

In the advanced levels, an even fewer number of suppliers work with other functions in the firm to find ways to augment the end-to-end processing that takes place. Designers from a supplier will work interactively over cyber-based systems with the manufacturer's engineers and production personnel to reduce design cycle times and build virtual inventory systems showing all goods in the end-to-end supply chain. In Level 4 these suppliers even work with the firm and its sales function to build revenues having meaning for both parties. Joint calls are conducted and joint investments are made in mutual assets, giving the relationship a distinct advantage in the eyes of the end customer or consumer group.

In Level 5, the most important of those suppliers are directly linked in the network connectivity so crucial to gaining the highest ground in an industry competition.

Among the new metrics that find their way into evaluating and guiding this function, the following can be listed:

- Reduction in supplier base — the number of preferred suppliers vs. total supply base, often tracked by major category of buy. The concept is that as fewer sources have larger volumes, prices are better and the purchasing staff has more time to devote to strategy.
- Supplier use index — getting to the number and percentage of suppliers that account for 80% of the firm's total spend, while having direct impact on the buyer's efficiencies.
- Average processing time — measuring the time for requisition and approval, to drive an effort to improve process steps and ease the order placement process.
- Percent of purchase transaction made through purchasing cards — an early measure designed to move small transactions to a local decision and reducing the costs of purchase and reporting.
- Percent of items accessed via a Web site (catalogs, auctions, etc.) vs. total number of line items — another measure intended to reduce the time spent on routine actions and take advantage of automated process steps.
- Cost of transaction processing — measuring the cost of entering, processing, and completing an order, from submission of request to payment to supplier. This metric is used to move a firm from a heavy reliance on telephone, mail, fax, and EDI ordering to some form of electronic processing.
- Percent of purchase orders sent without intervention — determining the number of orders that are submitted without intervention to determine if the approval process has been simplified and free of errors.
- Time spent on purchasing/tactical activities vs. strategic effort — a difficult measure, but one intended to drive the function toward automation of routine effort so more time is spent on strategic sourcing and working with the supply base in advanced efforts.

This list is by no means definitive or absolute. A firm must determine the role of this function in each level of its transition and then establish metrics that bring the intended focus to the function. At all times, the cooperation of this function with internal partners must be enhanced as a precursor to enhancing external relations with key customers, a group rarely accessed by members of the purchasing community.

All of the other functions must travel through the same transition as traditional measures that make sense are preserved and new ones that help the supply chain effort are merged to move the organization to the advanced levels of progress. This is best accomplished as the functional leaders develop a firm grasp of the supply chain concepts and suggest their own metrics. Logistics managers, for example, typically begin by breaking their costs into categories, including: transportation, warehousing, inventory, and administration. Metrics can then be brought to each area for improvement. In a typical advanced situation, total logistics cost will go from 14 to 16% of revenues (7.0 to 8.0% for transportation, 3.0% for warehousing, 3.0% for inventory, and 1 to 2% for administration), to 6 to 8% (3% for transportation, 1 to 2% for warehousing, 1 to 2% for inventory, and 1% for administration).

What Really Matters Comes from an External Focus on Customers

If we believe future competition will be between networks, then only total value across the extended enterprise end-to-end system matters. That has to be from a customer or end consumer view. Behind any new set of supply chain measures there must be some fundamental premises for guidance. The best reasons for inducing a company to adopt new metrics include:

- To improve profitability by bringing focus to intra-enterprise process steps that would not be improved with the typical silo-based thinking permeating most early-level firms. That focus must be end-to-end across the full supply chain that results in satisfaction of targeted markets, customers, and end consumers. That chain is only complete when the goods and services are accepted and not returned. Traditional measures that favor efficiency remain important in this aspect, but are best augmented with such measures as forecast accuracy (typically delivered through the sales function), returns and allowances (a measure of what went wrong), and order cycle time (how long it takes to get the processing done).
- To reduce sales, purchasing, operating, logistics, and administration costs to a degree not achieved with normal business processing. These

benefits come from measurements that lead to pinpointing the root causes detracting from cost, quality, productivity, cycle time, and customer satisfaction. A significant amount of administration cost, for example, is tied up in doing reconciliation because of bad order entry, errors in communication, and so forth. A metric to track the perfect order — a system that establishes the criteria for excellence in the eyes of the customer — helps to make progress here. Invoice accuracy must be a part of that system. Performance to request date is another metric that generally helps reduce logistics costs for buyer and seller by forcing operations to have the right good ready at the point of need.

- To bring a level of satisfaction to customers not achieved through competing networks. This satisfaction must be measured and matched with need, using a valid segmentation system that assures the greatest satisfaction by the most desired customer groups. Eventually, that satisfaction must be delivered to consumers targeted by network partners. New metrics here begin with on-time delivery, order fill rate, and customer service performance. These will be difficult metrics to meet in the beginning, but they are crucial to the advanced levels of supply chain as network partners seek unprecedented levels of satisfaction with targeted customer groups.
- To move an equal focus to top-line improvement; i.e., bring the firm to an external perspective where revenue growth is as important as cost reduction. These revenues will derive through exceptional service and exceeding the demands and needs of customers in traditional and non-traditional market segments, particularly as adversarial relationships are turned into delighted customers. Metrics that capture stockouts and back orders, and over/short/damaged shipments will bring focus to problems that cannot be tolerated if the network wants to build new business. Conversely, excellence in those areas is a tool for gaining new revenues.

Using this reasoning as a framework, the supply chain leaders are wise to consult with a small group of trusted customers and probe for the exact measures that have the most meaning and will drive the desired performance. Leading firms move to the point where an advisory council of such customers or distributors is established. Periodic meetings occur to review results of the new measures and to amend them as conditions change.

Action Study — Consumer Products Company

A vice president of manufacturing and product supply chain for a major North American manufacturer of consumer products has led an impressive assault on supply chain. The firm, a leader in household consumables, has been working hard at supply chain improvement for almost a decade. Under the leadership of the VP and his associates, the North American operations have made considerable progress and we rate that business unit a Level 4 organization. Their progress offers an example of how to do metrics correctly.

As the North American product supply chain took form, six strategies were made a part of a supply chain effort intended to establish "leadership through change." The six strategies were segmented by three key strategic objectives:

- Driving growth included global and business partnerships and financial leadership.
- Funding growth included ultimate flexibility/perfect reliability and relentless business simplification.
- Becoming the best place to work included world class customer service and quality, and health and safety leadership.

Work to set up key initiatives and metrics that derived from an orientation around these strategies led to establishing five areas of focus with the actual measures used. Each area had a designated leader and action team to assure implementation.

- New Business — This area paid attention to developing new revenues, particularly through specialty packaging. Measures focused on on-time and complete deliveries, amount of co-packaging, lead time, reliability, special packs in-house, and total cost. Other metrics brought focus to new products and measurements of technology transfer, 100% case fill, and new product introductions.
- Customer Service — This area focused on SKU portfolio optimization, including cost of complexity, item profitability, and SKU reduction. It also covered improved service and measures of forecast error, schedule/demand adherence, cross-border sourcing, and case fill rate.
- Quality — This was a crucial area that was kept to assure no deterioration in product quality. Metrics included on-shelf quality rating, consumer complaints, 100% positive release, and product releases.

- Return on Capital — Traditional measures were used here, but were augmented with measures of asset utilization, cycle time reduction, integration of business processes, capacity management, inventory turns, inventory as a percent of sales, capital spending vs. capital savings, and alignment of performance indicators.
- Personal Leadership — This area might seem unusual for supply chain, but the firm made it an integral part of its improvement process. Metrics included shared goal setting, cross-functional assignments, diversity targets, reduced OSHA rates, people development, rewards and recognition, and labor relations.

Each of these areas and metrics was backed up with a specific action plan, complete with a designated leader, targets, timetable, assigned resources, estimated costs/benefits, and anticipated challenges and opportunities. Periodic review sessions, where team leaders report on progress, have verified the success of the efforts and, with the help of the CFO, have been tracked to the bottom line of the business unit's P&L statement.

Summary

Companies will head in the direction determined by what is being measured. People will put effort on those areas that directly affect their compensation. With these concepts in mind, a firm does well to supplement its supply chain vision and effort with a set of metrics that encourage and support the right behavior. Traditional measures will never be eliminated, particularly those that relate to financial performance. Some of these measures and reward systems detract from a customer-oriented supply chain effort.

What a firm must do is slowly integrate a set of new metrics into the fabric of its culture. These measures must be specific to the firm's intended objectives and apply to all internal functions. As functional resistance is eliminated and an understanding of total costs across the internal enterprise is achieved, the company can move externally. In this external environment, with the help of willing partners, an end-to-end set of measures can be established. The actual metrics will be different for each industry and the intended customer or consumer groups that need to be satisfied. Care should be taken to review the measures used with actual customers to verify the application is going to achieve the intended external results.

4 Mistake 3: Aversion to External Advice

The thesis developed thus far is that to ensure competitive advantage in a chosen marketplace, companies must complete a supply chain evolution. They first need to bring forward the best elements of their continuous improvement efforts and merge them with an end-to-end focus on all process steps, and then begin a drive to optimize their internal operations. With a measure of success, the funding will be found to continue the effort and the firm can proceed to form information-intensive networks with its external trading partners. A compelling vision must then be developed that propels the organization into the external perspective. That vision must spell out the future state that will bring distinction to the firm and its allies in view of the intended customers and end consumers.

With a focused collaborative effort and a clear set of metrics that bring rewards for the intended results, the network will eventually migrate into a value chain constellation that is preferred by the intended end consumers. Once that happens, the mutually allied partners can strive for the ultimate supply chain differentiator–full network connectivity. To understand one of the major difficulties in the transformation being espoused, we will consider in this chapter the part of the evolution that takes the firm across the cultural barrier standing between Levels 2 and 3. As we do, we will see that the next mistake is being averse to accepting external advice.

This is a crucial problem as most firms making a successful progression to Levels 3 through 5 have learned — they cannot make the journey by themselves. These leaders favor the latest feature of advanced supply chain management — collaboration with trusted allies in their extended enterprise network. This chapter will also take an in-depth look at the fundamentals of collaboration as we begin setting the stage for explaining how collaboration and technology can combine to resolve the mistakes haunting supply chain. A further elaboration will be presented in Chapter 9 where we consider the lack of gaining the full impact from collaboration.

Supply Chain Is a Rich Vein of Business Opportunity

To begin, the critical importance of supply chain as the tool for attaining the next level of process improvement in a business must be re-emphasized. Of all the improvement techniques being considered today, none offers the range of potential achievement as an end-to-end focus on supply chain. Cost reduction potential peaks under this umbrella effort, as all process steps are brought under scrutiny for possible improvement. Operating issues across the full network are reviewed and most companies find many practices are not as effective as they could be. An intense effort to find the best possible practices becomes the key to bringing solutions to most of those issues. And other opportunities are generated that would not be addressed in normal relationships, as traditional supply chain partners find a wealth of new areas of interaction to consider for simplification and improvement. These are accomplishments that demand cross-area cooperation.

As a firm considers this added-value opportunity, it typically starts by viewing the supply chain in a traditional manner — as a linear, point-to-point, and unidirectional arrangement. That means the company focuses on optimizing the results close to or within its four walls. A manufacturer, for example, will consider the linear processes from supply to its plants, products through distribution, and deliveries to retail stores. A retailer will look at supplies from manufacturers through distribution centers, sales at stores, and returns to distribution. A healthcare provider will study shipments from suppliers through distributors to the healthcare facility.

Exhibit 4.1 illustrates such a linear flow. Firms approaching supply chain will try to manage and improve internal operations with little regard for the total network requirements. In spite of this limitation, progress will be made. Information sharing will improve, albeit reluctantly, between adjacent chain members. Suppliers will provide lower prices for larger volumes. Distributors will re-define their roles and add extra values. Transportation costs will decline. Some customers will work with a manufacturer to improve forecasting accuracy and reduce inventory needs.

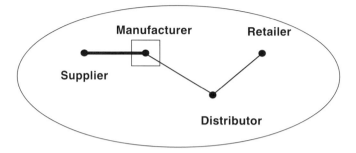

Exhibit 4.1 Traditional supply chain.

As further progress is made, and an infrastructure is attached to the effort, the focus on supply chain and cost reduction generally reaches a crossroad where the firm must decide to continue driving for internal excellence or begin to partner with external allies to get to Level 3 and beyond. At this point, the firm encounters the next dilemma interfering with smooth progress — the refusal to accept external advice. A strong cultural barrier is encountered that stresses the need to keep the continued drive for optimization inside the organization.

The position being put forth should be clarified. All firms will look at outside information. The problem comes in using the information (and the results of applications) and allowing external partners to become a part of the actual improvement effort. Every company today sends people to seminars and meetings on supply chain. They have people attend industry-sponsored symposiums. Books and articles abound and are read thoroughly. But a typical low-level firm will insist that the actual development of its supply chain effort be done with its own people.

To appreciate this dilemma, consider a conversation I had with a senior executive running the largest business unit for a Fortune 50 company. This business unit had achieved a good measure of improvement with its effort, but had reached the point of diminishing returns, and this executive was seeking counsel. When I suggested that further progress with his supply chain effort would require some external advice, he responded, "If I have to bring in outsiders, I'll fire the people I have." When I asked why he would do that, he gave a further reply, "because it would be an indication that I didn't have the best people working for me."

This individual is not atypical of the leaders encountered as further progress is sought. The general impression is that the internal effort is bogged down because the people are not giving enough effort to the tasks. That is not the root problem. Rather, it is typically a case where there is a need for an infusion of new ideas and external thinking. Progress with the individual cited began when we were able to prove that the lack of collaboration within his own structure was inhibiting further advancement. A few pilots later, with some carefully selected external partners, proved there was a wealth of better practices available if his people could combine their best thinking with that of partners who had found improvements in areas not under the internal scrutiny.

General Motors, a firm decidedly hampered by its historical and cultural inability to accept external advice or to allow outsiders into its inner workings, offers an example of a company that has discovered the value of overcoming the aversion problem. GM, whose ability to integrate its suppliers into a collaborative network helping the company to design cars has improved dramatically, discovered it "can save more than six months in overall design and production time. GM asks manufacturers of components such as seats, dashboards, and fuel-injection components to provide input on automobile designs very early in the conceptual phase, which lets the automaker find ways to reduce manufacturing costs by better accommodating suppliers' products" (Watson, 2001, p. 1).

In fairness, some firms believe if they have made good progress they must be ahead of every competitor, and bringing in outside help will only open the doors to letting the competitors have access to their secrets. A modicum of this concern is justified, but with so much general knowledge being spread so rapidly these days, the lead is often temporary. The better course of action is to solicit external advice, in a collaborative manner, from a few trusted resources so the next levels of progress can be achieved. If there are areas of proprietary information that would cause a competitive problem if leaked to the outside world, they should be clearly defined in the beginning of the collaboration and kept outside of the cooperative actions that take place. We would hardly expect Coca-Cola to share its formula with teams working on transportation improvement.

Supply Chain Requires Internal and External Cooperation

As a reminder, cooperation has to be a central ingredient in supply chain. There is no better way to encourage such a condition than through focused collaboration. But what is this concept all about? It has been termed the key driver of Wave 2 of the cyber-based Internet revolution. It has been called the cure for the NASDAQ slide. It has been dubbed the new way to find hidden savings in extended enterprise relationships. It has been given a name — collaborative commerce. But wait a minute! Collaboration is not a new concept. Smart people and companies have been collaborating with suppliers and customers since they cashed their first paychecks or bartered with chickens.

What has created all the interest in the marketplace and what has thrust collaboration to the forefront of business practices is not the newness of the idea. It is the feasibility to use collaborative techniques made possible through the medium of the Internet. That opportunity establishes what will become the new methodology for conducting business commerce — the ***networked supply chain***. It will be this methodology that determines which network dominates future markets and industries.

Through the Internet, collaboration takes on new and enhanced meaning as it now becomes possible to share data — drawings, designs, consumption information, new technologies, artistic renderings, promotional information, changes, approvals, and so forth — in an accurate and instantaneous environment. Business never had it so good! Inter-enterprise, networked business processes are now a real possibility by virtue of having the ultrafast, ubiquitous medium that was missing in the old economy. And collaboration among partners in this new environment, within and without the organization, is destined to be the enabling technique in the transformation.

The transformation is not in the way companies cooperate together for mutual benefit. That's old news. Some do it well; others do it poorly. It is in the time frames that are dramatically reduced, the accuracy of communications, the wealth of information that can be shared, and the elimination of nonvalue adding time making corrections or awaiting approvals. The advantages come from sharing of best practices, applying new ideas and concepts before others discover their value, speeding the time from concept to commercial acceptance, and increasing the probabilities of success in the marketplace by offering what consumers really want.

Collaborative commerce is the way in which enterprises in an end-to-end value chain interact electronically to plan, design, buy, build, sell, distribute, and support the goods and services that end-users consume. Through mutual efforts, collaboration becomes the design and implementation of processes and technology to enable an enterprise to work more closely with external partners throughout its value chain network. Such an effort begins, like all other elements of supply chain, inside the firm and progresses with the help of trusted external allies.

A few critical questions will help a firm determine if it is ready for collaboration on both fronts:

1. Are you sharing information, either internally or externally, on such processes as sales forecasts, inventory management, order processing, scheduling and planning, warehousing and shipping, and customer retention, for maximum benefit? Are there missing pieces of data in those areas that pertain to efficient processing?
2. Are you utilizing Internet technologies to improve process steps in your supply chain network? Have you applied cyber-based techniques for best advantage in the appropriate areas of supply chain optimization — order entry, order management, design and development, cycle time reduction, logistics improvement, etc.?
3. Does cooperative planning exist within your supply network? Have you and your allies taken advantage of the investments made in enterprise-wide resource planning? Is ERP functional and effective within your firm?
4. Is your supply network utilizing one balanced scorecard to measure improvements? Have external partners helped you develop that balanced scorecard?
5. Have you raised the level of trust within your organization and with external partners by sharing information critical to more efficient processing?

The answers to these questions will help a firm determine if it has begun to collaborate or not. If the basics are not in place, then the advanced opportunities will never be met. The process begins inside the organization as the various business units and functional departments start to put aside old animosities and indifferences and begin to share process improvements to streamline the internal portion of the supply chain. When the internal constituents have shown they can cooperate with each other, the logical move is to work with a few close allies on focused efforts. Now the firm faces the reality that the future is not in its hands, but in the hands of the network of which it is just a single unit.

Supply Chains Are Becoming More Complicated

Nothing will drive this advancement more than an appreciation for what has been happening to the supply chain at most companies. There was a time when the simplicity of the model described in Exhibit 4.1 would cover the process steps in a supply chain. That time is long gone. The linear model has been replaced with a complicated circular, multi-dimensional model. Exhibit 4.2 comes closer to describing today's conditions as the supply chain evolves into a collaborative network.

On the supply side, there are not just raw material suppliers. As firms progress their expertise and look at core competencies, they develop component suppliers to make things better than can be done internally. Contract manufacturers are used for parts and products better done by others. Design partners are allowed access to the research and development function to aid in reducing the cycle time for product introduction and to bring the best possible characteristics to new products and services. Several manufacturers may be linked together in a consortium that includes internal and external partners for the manufacture of an airplane, ship, or complicated medical device.

Multiple distributors can be used for complete coverage of a designated marketplace — to get food to all parts of a state, for example. In the high-technology market, resellers can be used to move product to a host of retailers or corporate customers. Hewlett-Packard sells a large percentage of its printers and other equipment directly to large corporate accounts and through such resellers as Best Buy and Computer Discount Warehouse. And multiple consumer groups could be targeted for consumption of the products. Along the bottom of the chart are just a few of the enablers who will be used to improve the delivery of the goods. It has become a very complicated field.

Since the business customer and end consumer have become the final determiners of what will be taken out of the collaborative network that has emerged, the new game becomes a function of the inter-enterprise cooperation that takes place to satisfy those constituents. That introduces challenges and requires modifications to historical organizational boundaries. The wise firm is hard at work determining the implications of this evolution, and is working with key constituents to design the network of choice that will satisfy targeted consumer groups. Most importantly, these networks are planning together the technical support needed to facilitate and enhance the increasingly complex business processes. Further planning is underway to develop the business models that will take advantage of the best technologies and applications. A fusion of business practices, information technology, and supply chain strategies will be needed to win this new contest.

Several specifics become a part of successful collaboration. First, the trading partners must work together, sharing what each has learned, to design and implement improved processes and install new technologies

46 ■ The Supply Chain Manager's Problem-Solver

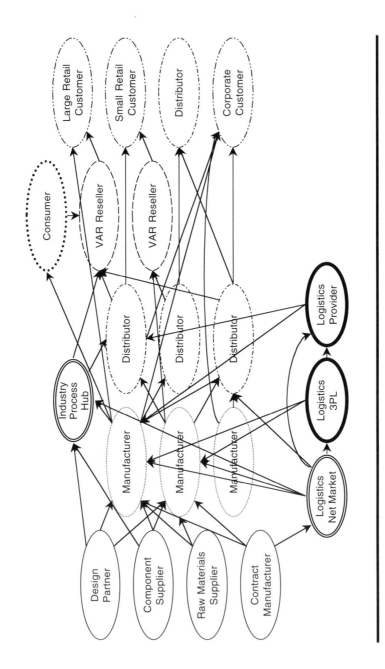

Exhibit 4.2 Supply chains are becoming collaborative networks.

that will provide greater efficiencies across the network. Second, collaboration can start with the simple sharing of information between these partners, but it must progress to an environment where several tiers and multiple functions in the value chain are collaborating. These interactions must be enhanced with access to real-time information. Third, progress will be best when a nucleus firm takes a central position in the effort. By nucleus firm, we mean a company that has the scale or brand equity that will drive the collaborative vision. Firms such as DuPont, Ford, McDonalds, Kraft Foods, and Boeing fit this role. They can form the network more easily than smaller entities and they have the recognized branding that can be enhanced with a collaborative effort.

A New Value Perspective Will Emerge

With these specifics understood, the collaborating partners are ready to take the best elements of their mutual supply chain efforts and combine them into a new value perspective. That means they begin working together to design network solutions. Exhibit 4.3 illustrates what can emerge.

In the center of the network will be the manufacturing services that are required to create the goods demanded by the customers and consumers. That is a place for a car, clothing, cosmetic, pharmaceutical, appliance, computer, or other nucleus manufacturer. It could also be the position of a service provider such as a bank, hospital, or travel agency, surrounded by its important constituents. On the left of the central position will be the supplier services needed by the nucleus firm. This could include other manufacturers, contract manufacturers, and traditional suppliers. On the right will be the customer services demanded by the business customers, distributors and resellers, retailers, and consumers. At the top will be those allies vital to product design and life-cycle services — the design partners so important for successful product introductions. At the bottom will be fulfillment services — the partners that enable the supply chain, including such collaborators as external logistics providers.

Together these constituents begin bringing focus to how they can share information, techniques, methodologies, software, practices, applications, and so forth to enhance the total processing from end to end of the chain. With data on what the total costs and values are across this extended enterprise, a new value perspective is established. Through mutual effort, propositions are created that add profits for the constituents and distinguish the network in the eyes of the intended customers and consumers.

48 ■ The Supply Chain Manager's Problem-Solver

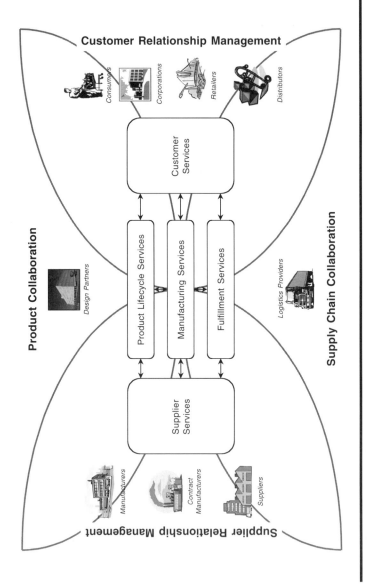

Exhibit 4.3 Supply chain value perspective.

For the cooperating constituents, the time to market for new products will be reduced. Assets will be better utilized across the network. Consumer data will be shared, rather than purchased, from external aggregators of such information. Joint customer presentations will be made to get new revenues for the network. Nonvalue adding process steps will be eliminated. Reliance on inventory will be lessened, as all inventories are visible and online. Capital expenditures will be jointly planned with expenditures and risks shared. The cash-to-cash cycle will be dramatically lower. Order cycle times will be at industry-low levels.

For the business customers and consumers, service levels will rise dramatically with the highest possible fill rates and on-time deliveries, with the least amount of back orders and returns. What is really in demand will be matched with what is available in the supply chain. Out-of-stocks will be at very low levels, if not banished. Customer service will be matched with need, through the option of choice — 800-number, Internet access, fax, or mail. Online visibility will allow for the diversion of product flow to the point of need. Obsolescence of goods in the system will disappear.

In brief, all of the potential benefits of supply chain will synthesize into a supply chain value perspective that has meaning for all constituents. The key to gaining this perspective can start many ways, but the most successful from first-hand experience is by setting a nucleus firm in the center of the collaborative effort. This firm then extends the effort with willing partners in a way that they all benefit from the emerging value perspective.

The Nucleus Firm Will Define the Starting Position

Exhibit 4.4 is an illustration placing the nucleus firm in the center of the new value perspective. Consider a network with a major industry player assuming that position — General Motors, Sara Lee, Coca-Cola or Pepsi, Citicorp, Boeing, Johnson & Johnson, etc. This type of firm has the scale and position to assemble all the players listed around its environment. Together, these constituents draw up the process map that describes their relationships in detail. Beginning with one or two of the most trusted partners, the nucleus firm begins diagnosing where improvements can be brought to those process steps.

On the buy side (the usual starting position), the firm works with all manner of important suppliers to find the means to optimize processing. Buying and selling portals are considered to leverage a larger buy and find additional sources of materials and supplies. Auctions are tested. Balanced scorecard techniques that evaluate the total cost of ownership are implemented. Electronic procurement techniques are introduced to reduce the transaction costs for both parties and selective outsourcing is considered. In later stages of collaboration, the firm returns to this side to find the means of cooperating together to build new revenues, often in nontraditional market sectors.

Working with one large manufacturer of carbonated beverages, we were able to reduce the total cost of production by a third, as the best external source (contract bottler, for example) was used for each territory sector and necessary raw materials were bought in an aggregated manner by the nucleus firm and its contracted manufacturers. We also found the way to work together with these contractors on selling one of the largest private-label deals to one of the largest retailers in the U.S.

On the product side, the nucleus firm expands its collaboration with design partners and complementers, firms needed to complete a design or product introduction. Using the experience gained with the suppliers, the technique is extended, often with some of the key suppliers in attendance. Product planning and strategy become a joint effort, as does sales product configuration. Requirements management becomes a shared interest as the parties work to smooth the flow of parts and supplies into the design center and later into manufacturing. Teams work on product content, necessary resources, and schedule optimization, helping all parties get to lower costs and higher efficiencies.

In advanced efforts, electronic bid, proposal, response, and purchasing techniques are introduced. Technology selection and implementation is done jointly. Hosting and application service provision become a team effort. Ford was very successful with this overall approach when it introduced the Mondeo model to its European buyers in a record-setting 16 months — unprecedented in an industry that typically works with a 48- to 60-month cycle. The secret was allowing the component suppliers and external designers to have direct access to the design process. That means the partners were connected via their computer-aided design and manufacturing systems into Ford's central design system. Suppliers were able to make direct design contributions. Plans and specifications were reviewed online, approvals were accelerated, and changes were handled in a matter of days and not months.

On the enabler side, the nucleus firm and selected network partners work with external firms to better handle logistics, credit, regulatory compliance, and payments. Particular progress has been made here as firms are forming virtual logistics networks that seek out, from hundreds of potential transportation companies, available capacity in planes, trucks, railroad cars, and ships. (An elaboration on this subject will be presented in Chapter 8.) They also cooperate on architecture and design of new systems, planning for promotions and distribution, and how to optimize use of joint assets. Order and inventory management become key issues here as these enablers are solicited to find ways to get to the next level of improvement.

On the sell side, the nucleus firm now considers the many channel options and develops solutions with each of its constituent customers. A DuPont will work with other corporations to find more efficient ways to send chemicals and products such as Nylon, Teflon, and Spandex to firms needing these raw materials for further processing. Kraft Food will work with food distributors Fleming or Supervalu to distribute its products to small grocers, and to major retailers Kroger and Safeway for direct-store distribution to their retail outlets. Together, these firms look at sharing customer segmentation data and analyses. They design portals and network exchanges to better facilitate bidding and order entry. Call center design becomes a joint effort.

For all of its positive aspects, there are impediments to successful collaboration. The process requires credible, accurate, and timely data. There is also a need to understand the real costs of business activities. These requirements often restrict a firm from participating in collaboration, particularly if they have not sufficiently completed levels 1 and 2 of their supply chain

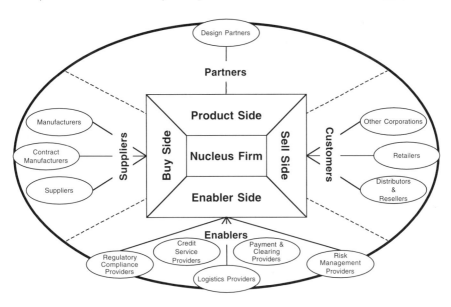

Exhibit 4.4 Adding detail to collaborative commerce; be wary of the impediments.

progression. Collaboration proceeds best when parties of equal capability are sharing their best ideas. Having to educate a slower or less capable partner takes more time than the typical patience displayed in today's dynamic business environment.

Sufficient skilled people must be available to do the analysis and work on the requisite implementation teams that result from collaborative efforts. That is particularly true of a traditionally strapped function — information technology. With technology being such an integral part of the solutions that will be developed, a firm is wise not to enter into collaboration without the available time from those who will design the future systems.

Action Study — Moen, Inc.

Jeffrey Svoboda, president of Moen, Inc., a North Olmsted, Ohio manufacturer of plumbing fixtures, has used the elements of advanced supply chain to take advantage of recent changes in that market, as consumers have begun demanding upscale styling and features in what used to be a rather mundane business. Svoboda terms it a "9-to-5" strategy through which the 54-year-old company turns out new designs as fast as consumer preferences change. Beginning the process improvement, his instructions were to cut out the fat in Moen's systems, speed new products through the supply chain, reduce inventories, and free up cash for new investments. The key to this transition was using the Internet. A particular success came in using the Web to design new and innovative products in record time.

By collaborating on designs with key suppliers over an extranet connection, a new Moen product can go from the design stage to a store shelf in 16 months, down from the traditional 24 months. "The time savings makes it possible for Moen's 50 engineers to work on 3 times as many projects, and introduce 5 to 15 fashion (faucet) lines a year" (Keenan, 2001, p. EB 17). With more products reaching the market in less time, Moen has seen a 17% increase in annual sales and the company has vaulted from the third position in the industry to a virtual tie with Delta Faucet Company for the leading position.

Beginning in 1997, Moen's technology chief, Tim Baker, and his Internet Program Office team set priorities for using the Web to achieve an advantage. Deciding one of the first moves had to be to streamline its product development cycle, the team set about improving the process steps in that function. Previously, the firm would take 6 to 8 weeks to design a new faucet. The design would be captured on a compact disk and sent to qualified suppliers in 14 countries. The suppliers might find they could meet the specifications or not, but often they would make changes, prepare a new CD, and return them to Moen. These changes would be combined with those received from other suppliers to get a complete design. If some of the changes were incompatible, the process would have to start over. The total time was not acceptable for the new market thrust.

Carefully spending $1.5 million, Baker hired software developers to work with his team on internal work and external features like an online design room that allows customers to mix and match shower features. Now the input from the actual consumers is used as source data for the new designs. With this information, the designs were customized to the market demands. Beginning in 1998, Moen started "sending electronic files of the newest product designs by e-mail. A few months later, it launched Project Net, an online site where Moen can share digital designs simultaneously with suppliers worldwide. Every supplier can make changes immediately. Moen consolidates all design changes into a master Web file. That way, design problems are discovered instantly and adjustments can be made just as fast, cutting the time it takes to lock in a final design to three days" (Keenan, 2001, p. EB 20).

In a later move, the firm decided to improve the order management process, which was heavily dependent on mail and fax communications. In October 2000, the firm launched Supply Net, an extension of its extranet, which allows parts suppliers to check the status of Moen's orders online. "Every time Moen changes an order, the supplier receives an e-mail. If a supplier can't fill an order in time, it can alert Moen right away so the faucet maker can search elsewhere for the part. Today, the 40 key suppliers who make 80 percent of the parts that Moen buys use Supply Net. The result: the company has shaved three million dollars, or about 6 percent off its raw material and work-in-progress inventory since October" (Keenan, 2001, p. EB 20).

With these successes, Baker's team has turned its attention to Customer Net, an ambitious effort to wire the firm's wholesalers, which account for half of the company's business. Using another calculated approach, the firm brought a select group online at the end of 2001, as they extended their extranet toward the consumer.

Summary

The concepts presented here will be revisited in Chapter 9 as we consider why collaboration is not more extensive, given its obvious benefits. For now, let us understand that collaboration adds the kind of detailed analysis and finding of solutions that elude a firm determined not to accept external advice. It begins by erasing the cultural barrier that schools a firm and its people to think they have a lock on all the good ideas. With so much happening so rapidly in today's economy, the path to the future is best traversed with the help of willing trading partners who can share their best ideas with yours so the total becomes an enhanced network providing the kind of services sought by today's consumers.

5 Mistake 4: Focusing Only on the Bottom Line

No amount of effort or exhortation is going to distract a business organization from its focus on constantly improving the bottom line. Net profits are going to be the driving force behind corporate existence for a long time. Overemphasis on this objective, however, can become another inhibitor to supply chain progress. On the one hand, as a firm seeks to increase its profits by finding the means to reduce costs, supply chain becomes a valuable tool. On the other hand, if the firm continues a single-minded drive for internal improvement, it puts itself in a position where other companies become less willing, or even reluctant, to partner with the firm. With so many firms pursuing internal excellence, there are only so many resources that can be appropriated for improvement efforts. Eventually, sharing such resources with other companies only happens when there is some form of mutual reward or a superordinate objective to achieve. Advanced supply chain management efforts are characterized by a shift in focus from the bottom line (cost reduction) to the top line (profitable new revenues).

In-between those two lines are a myriad of areas that impact the well-being of the firm. As companies reach the point of diminishing returns from their endless cost-saving efforts, a new look pervades the firm. Now the company focuses on the top of the P&L statement and moves through each category to determine how improvements can be made, expecting profits to be enhanced. Reducing returns takes on new meaning as firms work with key customers to get at root causes for problems and make them go away forever. Working with key suppliers to find features that delight customers, cutting cycle times for new product introductions, and moving from a reliance on sales forecasts to online access to actual consumption become new initiatives. Being able to see what is in the supply chain pipeline becomes extremely important. Outsourcing entire functions to more competent partners is seriously considered.

Leading firms realize a passion for having the lowest cost position is fine in a mature, static market. There is not much else to pursue as market shares are fairly stable and customers are generally looking only for price differences. In more dynamic markets, this strategy overlooks the fact that what customers and consumers want is innovation, features, flexibility, quick response, good deliveries, solutions, and an easy way to conduct business. These characteristics come from an effective network, not a super-efficient individual firm.

Coach, Inc. is an example of a firm dedicated not just to having extremely competitive costs, but producing what their targeted upscale customer wants. This firm has been hard at work redesigning its entire supply chain network, from foreign and domestic suppliers, through multiple distribution channels, to its own retail stores and factory outlets, as well as other retailers and its specialty units. The strategy is to stay on top of fashion trends and build more than half of its annual sales from new product introductions. Implementing this strategy requires making the suppliers, mostly Asian, a part of the design and development process, and having a distribution system that makes certain the items in demand are at the point of purchase when required. Interestingly, the emphasis in their transformation was not on new sales. That would come if the changes were successful. The emphasis was on creating the most effective system for new product introduction on a global basis.

In this chapter, we will consider some ideas and techniques we see in leading firms that have moved at least a part of the supply chain effort off the traditional, single focus on cost reduction. We will illustrate how the new business game revolves around attention to the top line and the intermediate processes that lead to a strong bottom line.

The New Game Starts with the Consumer

In their book, *The Myth of Excellence,* Fred Crawford and Ryan Mathews have brought attention to the fact that it is impossible for a company to be excellent at every important aspect of its business, particularly those having meaning for today's consumers. For these authors, universal commercial excellence is a myth (Crawford and Mathews, 2001, p. 3). The authors build a strong argument throughout their book for being capable at the key consumer elements (price, service, access, experience, and product) and excelling at those that make a difference for the intended consumer group. This capability requires cooperation across the network and attention to the key things that differentiate the network, not just having the lowest costs and prices.

With the authors' comments in mind, we reflected on our experiences with changing consumer patterns and concluded that all supply chains, no matter how far upstream or how much related to a commodity situation, should be focused on the end consumer. As a particular group is targeted, the requirement quickly comes into focus on how best to satisfy that group. Now the network can bring attention to the consumer elements that make a difference. Wal-Mart may target all people interested in a large assortment of reasonably well-known products at the very best prices, while Target may focus on a slightly more upscale audience. Dell Computer might target the average home user with a focus on price, while Hewlett-Packard could target the industrial user through better service. These decisions not only affect the product offering, but the partners with whom the firm needs to rely for the inevitable demand for good selection and access.

Another factor comes into play as well. In the current marketing arena, an important trend is providing the targeted consumer with at least a feeling of customization; a sense that the product selected somehow has personal meaning. Such a condition requires a modicum of build-to-order in the manufacturing process and a collaborative trading network that assures delivery of what is wanted in the most efficient manner. Anything short will send buyers to an alternative source. Hewlett-Packard has gained a reputation for its ability to ship generic communication peripherals, such as printers, to distribution centers where they are localized upon receipt of a particular customer order. More than ever, firms are faced with the reality that price is just one factor, and often not the most important, in securing consumer loyalty.

Today, it is the ability to remain flexible in the face of changing consumer patterns and market conditions that is decisive for long-term success. With the appearance of mass customization came the need for collaborative supply chains, with the information flows not under the control of a single entity. Multiple parties, working toward a condition of extended enterprise optimization, now require coordination. Sharing responsibility for process steps designed to please the consumer results in meeting operational goals in the current business environment.

Supply chain networks are integrating product information such as buying patterns, available-to-promise inventories, inventory status, point-of-sale consumption, and so forth. Across the network, partners work with consumer demand (cash register receipts, distribution center releases, etc.) to determine what is selling and what is not. Intelligent business decisions now depend on viewing this data, which matches what is happening in the external marketplace with what is capable from the now efficient internal system of response. The message content no longer relates only to historical events, but is related toward what is happening and what has to happen to sustain satisfaction of the targeted consumer group. In short, visibility across the total network is a requirement for success, and that requires an external focus on all process step improvement and not lowest cost.

"Carrier Corp. is using data mining to profile online business customers and offer them cool deals on air conditioners and related products," according to Rick Whiting at *Information Week*. By using services provided by Web-Miner Inc., this manufacturer of heating, ventilating, and air conditioning equipment has turned Web visitors into buyers, "increasing per-visitor revenue from $1.47 to $37.42." Carrier, a business unit of United Technologies, has been selling products via the Web since 1999, but garnered only 3,500 units that year. In 2000, the firm "gave WebMiner a year's worth of sales data, plus a database of Web surfers who'd signed up for an online sweepstakes." WebMiner combined that information with third-party demographic data to develop profiles of Carrier's online consumers and went looking for ways to satisfy the Web buyers. Online sales had exceeded 7,000 units through August 2001 (Whiting, 2001, p. 55).

Carrier and other consumer goods and industrial goods firms are now refocusing their efforts toward the consumer. As they do, they discover the importance of working with the intermediate business customer. The same type of attention to consumers applies as a firm seeks to build sales with business customers. In this sector of supply chain, we encounter generally more sophisticated players adept at using collaboration and technology. But it requires more than just a single point of contact to derive the most benefit from the new focus.

Business Customer Interface Is No Longer a Sales-Only Function

With the emergence of the consumer as the dominant entity in business commerce came another need — to realize that it takes more than direct sales contact to meet buying needs. Where such interface was once limited to sales, marketing, and customer service personnel, the need for operational and support functions to also be involved with business customers has become an accepted manifesto. Sourcing, engineering, manufacturing, design, development, distribution, and other functions now require customer data to optimize internally, and to bring the most overall value to the desired customer. That means the market leaders are not only internally proficient, they are becoming masters of customer service supply chain management.

Paul Fortner, director of E-business at Toledo, Ohio-based building products company Owens Corning, has found one way to accomplish that objective. His firm has designed an initiative to build personalized, online relationships with its business customers. Using software from BroadVision, the firm has added richness and an element of customization to its basic product and service data by introducing content management tools that "consolidate Web pages into small groups of personalized portals. Visitors, employees, customers, and partners will be able to access the portal that contains the most relevant information." The idea is to deliver value to specific audiences, according to Fortner, providing the "right information to each builder. For example, information on wet-weather building materials will be provided to regions with heavy rainfall" (Colkin, 2001, p. 26).

As the functional interface expands, so does the opportunity to find new savings and the path to the desired revenues. Engineering comes into play to find innovation features for which the desired consumers will pay more. Manufacturing works on planning and scheduling with network partners to eliminate out-of-stock conditions and to be flexible in its response to changing consumer patterns. Logistics finds the way to get shipments to the right place, sometimes in lot sizes of one, to an individual address. Materials management reaches back to sourcing and the key suppliers to make certain the needed bills of materials are filled out completely so production is not delayed. Collaborating firms involve whichever function is necessary in an integrated manner to meet the needs of the targeted group while providing a sense of customization to the effort.

Take a look at General Motors: the automotive giant is hard at work with partners across its supply chain network to build new car sales. This firm has determined collaboration with its partners, by integrating them into the design process, can "save more than six months in overall design and production time" (Watson, 2001, p. 1). GM has been working with suppliers to provide input on new car designs very early in the conceptual phase. This effort has enabled the automaker to reduce costs, but also to get the desired car models to the showroom sooner. The entire automobile industry is collectively working to bring that time frame down to the 10-day car — from order to delivery.

In another middle-of-the-P&L statement effort, rather than relying on large inventories placed strategically in advance of actual consumer demand, network partners are finding the means to make their supply chains capable of responding in real time to actual sales so more sales can be garnered, rather than lost. By focusing on visibility across the end-to-end network, and not just cost reduction, for example, some networks are forging new frontiers in supply chain responsiveness. These leaders use visibility to enable them to analyze and react to the interaction between procurement from multiple key suppliers, shifting manufacturing across internal and external assets, and transferring goods and services to the final business customer and end consumer.

Consider Federal Express, a firm always looking for ways to improve its supply chain capability through better services. FedEx's latest feature has been dubbed Insight, a system enhancement that "adds a set of proactive notification capabilities that will alert customers to key logistics events," according to CIO Rob Carter. If a customer's package runs into trouble at customs, for example, FedEx will send a message to the customer, via e-mail, cell phone, or pager, with information on the problem and advice on how to work around the situation. "We're going to continue to raise the information bar like that," says Carter, "and give customers visibility into their shipping processes" (Karpinski, July 11, 2001, p. 1).

The Game Expands with Business Partners and Customers

As the supply chain firm and its partners begin to increase the functional interfaces and get more parties involved in the pursuit of optimization and new revenues, they can take advantage of competencies across the extended enterprise, and not just within the internal four walls. Now the game is extended with the help of trusted business partners and customers interested in mutually developing new revenues. These revenues can come from business customers as well as targeted consumer groups.

Amazon.com has been a front-page story for some time in the journals, magazines, and newspapers chronicling e-commerce events and the ups and downs of the dot-com world. What has been missed in most of those stories is how this digital leader has been hard at work taking advantage of its infrastructure and e-commerce capabilities to build supply chain alliances and enhance the top line. The firm has been striking deals with often larger organizations to become the order or delivery arm in specific networks.

Under a special arrangement, for example, Amazon will carry electronic products from Circuit City stores on its well-known Web site. This arrangement is the third between the familiar book marketer and major retailers. Toysrus.com and bookseller Borders Group have transferred management of major parts of their e-business infrastructures to Amazon. "Under the agreement, Amazon will get a part of the revenue for Circuit City products sold through the Amazon site. Amazon will process transactions and Circuit City will fulfill products and provide product-related customer service. The brick-and-mortar merchant also will accept returns on items picked up at its stores." Securities analyst Merrill Lynch estimates that Amazon will get an 8 to 10% revenue share from Circuit City. "Because Amazon bears few costs for transactions beyond credit card processing, the online retailer will get more than a 50 percent operating margin on the revenue from the Circuit City deal" (Kemp, 2001a, p. 1).

Other firms are looking for novel ways to raise top-line revenues in partnership with other supply chain constituents. Any avenue seems to be a fair venue as traditional players are developing innovative and successful features across their networks. Wal-Mart, always on the forefront of these events, is giving sales and inventory data to its key suppliers to smooth the flow of delivery to stores and point of need where the consumer is ready to make the buy. By matching what is being consumed at the store with what is available in the supply chain, out-of-stocks diminish and revenues keep going to the top line. This procedure dramatically lessens the dependence on sales forecasts, which can be reasonably accurate on an annual basis, but have little relevance to daily operations.

Nestle Corp. is working with grocer Sainsbury in England to plan promotions together, and to involve other supply chain constituents, so the ingredients, packaging, and labels are available to match the flow needed to make the promotion a success. Big Boy Restaurants International turned from Coca-Cola to PepsiCo Inc. after 65 years, "due in part to PepsiCo's promise to build an intranet that Big Boy can use for internal, and eventually external, communications. Pepsi and Big Boy plan to roll out the network tool to all 170 of the Big Boy restaurants" (Lewis, 2001, p. 1).

Collaborative Technology Becomes a Key Ingredient

As partners work closely and innovatively to seek coveted new sales, they need tools of implementation. Turning to the need for information and good communications, the trend is to augment new revenue efforts with collaborative technology. Michael Walsh, director of technical standards and systems for Procter & Gamble, directs a web of technology that spans the consumer products firm's global enterprise. "P&G's computer network links 900 factories and 17 product development centers in 73 countries" (Ante, 2001, p. 146).

Through this global network, the company is able to produce and market 300 leading brands, from detergents and cosmetics to coffee and oral care. Forced to find the means to cope with speed, quality, and efficiency across such a far-flung enterprise, P&G set about to automate its product development process using collaborative technology. Called the "Corporate Standards System" (CSS), software from MatrixOne was employed to let researchers browse a database containing 200,000 product designs to see if they already exist in another part of the firm and eliminate redundant efforts. Product design times have been trimmed by half, mainly by letting the extended developers come up with designs together on the Web, while enabling managers to measure progress against predetermined timetables. "New collaborative tools like CSS make global science easier to do," says Walsh (Ante, 2001, p. 146).

The tools are coming from a new generation of firms helping supply chain efforts. MatrixOne, Agile Software, Nistevo, and Logistics.com are new names in this arena. The tools are supplychain focused, easier to use than their predecessors, and focused on specific solutions needed by the network players. The Limited Inc. used to float bids to 35 trucking companies via telephone and facsimile, so the truckers could send back offers for transporting products throughout the clothier's system. Lacking the desired efficiency, the firm, through its shipping arm, Limited Logistics Services (LLS), turned to Web services provided by Logistics.com.

LLS now invites shippers "to bid on jobs and develops an optimal delivery plan by weighing factors such as traffic congestion and truck availability. Instead of LLS paying for pricey one-way trips, Logistics.com finds shipments from other LLS warehouses to fill trucks for return trips." The shippers can offer a good price based on the information they receive and take advantage of other loads in a manner that allows them to fill their trucks. "Thanks to the service, LLS has sped up its average delivery time by a week, increased the reliability of its shipping rates by more than ten percent, and cut its delivery costs by $1.2 million, or three percent this year" (Ante, 2001, p. 147).

Case Study — Office Depot

Office Depot affords us one more example of how firms are turning to collaboration and technology to enhance already good sales performance. Under the direction of Monica Luechtefeld, Chief of e-commerce, this office-supply company has been luring business customers to use what is generally considered a consumer-oriented firm for their business needs. Beginning in 1996, Leuchtefeld worked with the Massachusetts Institute of Technology to set up a Web purchasing site for the university. From that experience, she convinced senior management to incorporate "the online effort as the backbone of the company's supply chain" (Haddad, 2001, p. EB24).

Ms. Leuchtefeld has continued to work with major organizations, convincing them they can save money using a system that even technical novices could master. Bank of America, for example, allows employees to order supplies from their desktop computers, receiving rebates in exchange for online purchases. The bank now orders 85% of its office supplies through Office Depot's online store and saves millions of dollars in the process.

Office Depot uses one seamless network to track inventory and sales, whether the goods are online, inventoried in a store, or available through a catalog. They can manage inventories on a real-time basis and build a larger position in the market as more business firms use the Web site to get the goods they need quickly and efficiently. The company is going further to set up kiosks in its stores so customers can find what they need, even if they don't see it on the shelves. The regular sales force has not resisted this feature, but actually helps some customers use the Web access.

"Today, 40 percent of Office Depot's major customers are using the online network to buy everything from cherry conference-room tables to paper clips," reports *Business Week* writer Charles Haddad. "Office Depot's online unit booked $982 million in sales last year — nearly double that of its biggest competitor. Last year, the company's Internet sales grew 143 percent, compared with a 12 percent increase in overall revenue. This year, the company expects its online sales to rise 30 percent, to $1.5 billion, and contribute 14 percent of overall sales" (Haddad, 2001, p. EB22).

Summary

Companies are going to concentrate on the bottom line of their profit-and-loss statements for a long time, as the drive for profits is what sustains a business. Nevertheless, there must be a shift to how the firm can work with willing allies to enhance the process steps that affect everything from the top line to the bottom. As supply chains mature, we see a stronger emphasis emerging — end-to-end processing that takes place as partners work together to get the best values for business customers and end consumers.

That emphasis requires focusing on how any step in the sourcing, manufacturing, and delivery process can be improved so the customers and consumers will elect to buy from a particular network. The new game is building revenues, even when one or more constituents might have to bear a higher cost. Collaboration and technology are the tools of choice in this business transformation.

6 Mistake 5: Poor Customer Relationship Management

Once the firm moves from a cost reduction-only focus and begins looking at building top-line revenues, it enters the area of customer relationship management (CRM). With access to a wealth of information relating to specific customers, conventional wisdom says a company can work with its sales, marketing, and customer service personnel, and selected internal and external partners, to enhance revenues with the most desired customers. Here, again, a single-minded focus on getting a direct and positive impact to profit can inhibit the effort and preclude attainment of the intended objectives.

Although the CRM concept was born of using technology and information to enhance sales and satisfy customers, most efforts are characterized by an emphasis on reducing costs, particularly sales and service headcount. To date, the call center has shown the highest returns from CRM, primarily through cutting costs and better resource allocation, not from helping the customer. Rather than building the expected top-line growth, many efforts languish as they become a control mechanism over the sales force, sales personnel do not enter the necessary data into the program, or the information is not used for the intended purpose. In this chapter, we will investigate the reasons behind the general lack of positive results from what should be a very powerful extension of a supply chain effort, and make some suggestions for getting more out of such initiatives.

The Problem Begins with Understanding and Purpose

Let us start by recognizing that using technology to increase sales is a poorly understood concept. Further, automating processes intended to generate new sales, without doing something to make things better for the customer or improving life for the sales and service personnel, generally does not work. According to Barton Goldberg, author of *The Guide to CRM Automation*, "If you don't have a good customer-facing process and you automate it, then you've just automated the problem" (Maselli, 2001, p. 40). As a result of these and other complications, many of the applications fail to reach the intended level of results. After-the-fact analyses of these events generally reveal there was poor up-front understanding of what exactly the firm expected to accomplish and results that were not related to the intended purposes.

Consider the dilemma faced by a major retailer — Saks, Inc. Denise Power, writing for *Executive Technology* magazine, reports on that firm's effort to come to grips with what could be accomplished with a CRM effort:

> It sounded like a good idea — invite vendors of customer relationship management software to parade their philosophies and solutions. The exercise yielded more questions than answers, however, and, in the end, Saks was left with the sobering realization that it had no solid definition of the term 'customer relationship management.' (Power, 2001b, p. 11)

"How then," Power questioned, "could the 350-store retailer, with direct-mail and online businesses, draw a viable road map for proceeding with a CRM strategy?" The firm had appointed a manager of CRM, Mike Gardner, but he was just as puzzled, remarking, "We have to figure out what CRM is to us, what we think we can get out of it, and who is a party to it internally." He further reported the firm was "preparing a white paper containing an assessment and series of recommendations on how to move forward with CRM at the $6.5 billion company" (Power, 2001b, p. 11). Hopefully, that white paper will include a solid purpose behind the effort, or Gardner will discover, like many before him, that it becomes an expensive investment in technology with no solid objectives.

There was another problem to be resolved by the retailer. Confused by the many definitions put forward for the meaning of CRM, as each functional area presented a different view, Saks decided it had to start at the beginning and come up with a definition that had meaning for the firm and would motivate support within the organization. Gardner is well advised to take this tact, as most companies jump into CRM without a clear understanding or purposeful road map. After much study in this area, I have concluded there is little in-depth understanding of the inherent requirements and now have little wonder why some industry analysts project a failure rate for CRM implementation at 65%, with some going as high as 80% by 2003.

Definitions of CRM will be presented in a following section, but first a firm must make certain there is no misunderstanding on purpose. CRM must be intended to enhance customer relationships, not to create bottom-line savings for the firm. The top- and bottom-line improvements will come when the firm takes care of its customers.

CRM Extends with Strategy and Technology Application

With a true customer-centric orientation built into the firm's purpose, and a clear understanding of what is behind the CRM effort, a company can move forward, but it will encounter the next problems — developing a strategy and introducing the enabling technology. A lack of cohesive CRM strategies impedes progress in virtually all industries. Companies simply do not use the information and tools available in an effective manner to drive more revenues from new and existing customers. At the core of the problem, the move from a product-centric orientation to a true customer-centric transformation is harder to achieve than originally anticipated.

It all sounded good, but it was more difficult that expected. As supply chain efforts matured and the internal house seemed to be in order, many companies turned their attention to the top line. "We're going to concentrate on what the customer wants and needs and build our future around those perspectives," many executives extolled. That call was quickly followed by an urgent desire to mine the burgeoning databases to find information — histories, trends, patterns, etc. — that would help the marketing, sales, and customer service groups to solve problems and build new sales. Data on past promotions would be used, for example, as a decision support tool to plan more effective, future promotions. Overlooked in this thinking was the fact that such an effort requires cooperation from trading partners, many of which were not consulted or did not have the technical capability to fully collaborate.

Undaunted by the lack of results, most firms moved aggressively forward, attempting to use information as the key to building new sales. But hold on a minute! This kind of effort requires an extensive use of technology to get at information buried in the databases companies are creating as they collect every speck of data they can on customers — their buying habits, trends with their consumers, past histories, results of joint calls across the firm, and so forth. Our study of hundreds of firms engaged in supply chain management and its offspring, CRM, reveals that few know how to properly apply technology to something as sophisticated as culling the appropriate information from data and using it effectively to enhance the building of new revenues.

Furthermore, the effort requires taking the benefits of the technology to many groups unprepared to use the information in a way that will enhance sales performance. It is a rare firm that can anticipate the human reactions to such an effort and properly prepare for the inevitable pushback from sectors of the organization not used to relying on technology to enhance performance. That is particularly true when participants have to develop a discipline to enter the required data into the new software in a timely fashion. It is a requirement that is only met when the people involved see the purpose and a direct correlation to improvement in their specific area.

CRM, according to industry analyst, Gartner, Inc., is supposed to be "a customer-focused strategy aimed at anticipating, understanding and responding to the needs of an enterprise's current and perspective customers. The objective of a CRM strategy is to optimize profitability, revenue and customer satisfaction" (Gartner, 2001, p. 1). That is a great concept, and one that is hard to ignore as a firm finds it has enhanced its ability to deliver what most customers want from a supply chain. Unfortunately, it requires a strategy and the introduction of software and techniques to extract the information crucial to "anticipating, understanding, and responding to needs." It fails to answer the question, what is in it for the sales and service people who must gather that data and use it effectively with customers?

Consider the further elaboration by Gartner, Inc. on the inherent aspects of CRM. According to this expert analyst,

> CRM solutions consist of the hardware, software and services deployed to support customer-facing processes that enable organizations to analyze, manage and alter the relationship between external customers and the organization. In their most extensive implementations, the solutions enable the organization to capture customer data from across the enterprise, consolidate all internally and externally acquired customer-related data in an integrated data repository, analyze the consolidated data, distribute the results of the analysis to various constituents of the extended enterprise and use that information when dealing with the customer. (Gartner, 2001, p. 2)

The firm cites the lack of a clear, cohesive CRM strategy, as well as the lack of understanding of CRM as major inhibitors to market expansion and rate of adoption of what should have been a major supply chain improvement tool. We agree wholeheartedly, and would add that unless the strategy brings value to the implementers and the targeted customers, it is all a waste of time.

As the idea caught hold in business, however, a slew of software specialists appeared, offering the kind of technology solutions that would deliver the intended results. The software sales representatives, however, were way ahead of their clients in understanding what the applications were all about, how they would impact the organization, and what the inevitable human reaction would be to introducing technology to such staid and traditional functions as sales, marketing, and customer service. The result has been *software before business proposition* and a littered landscape with more failures than successes.

To move to the positive side and illustrate how you can get it right, let us consider the case of a firm selected by *Information Week* as its leading business user of technology — medical and surgical supply distributor, Owens & Minor Inc. This 119-year-old Glen Allen, VA firm delivers medical supplies to its healthcare customers through strategically located warehouses. At the heart of its supply chain and CRM strategy is the intent to provide excellent service along with critical information provided through its data warehouse and use of the Internet. The idea can be summed up as using IT technology to "bolster the efficiencies of its business — and that of its customers" (McGee, 2001, p. 39). The company has customers on one side that manufacture medical and surgical supplies and equipment, and healthcare providers, such as hospitals and health maintenance organizations, that need these supplies for their patients.

In one move to bring technology features to its services, O&M decided to provide manufacturers with information on the usage of their products so they could better manage inventories and logistics. On the other side, they wanted to provide the cash-squeezed healthcare providers with information to help them "better manage time and costs related to their supply chains without having to make significant IT investments of their own. We realized our difference (between competitors in the market) would be based on providing information," says CEO Gilmer Minor (McGee, 2001, p. 40).

The strategy extends to becoming a supply chain information intermediary in its business channel, including the manufacturers and providers that historically share little data with each other. Through a system dubbed "Wisdom," the firm analyzes supply chain information on both sides and relays valuable data to both constituents. For O&M manufacturers, Wisdom provides "analyses of product penetration, contract compliance, drop-ship activity, inventories at specific distribution centers, and historical use of products. Suppliers have real-time access to purchase order information and the most efficient transportation of goods" (McGee, 2001, p. 40).

The next iteration of Wisdom is intended to provide even more decision support information with purchase histories for all supplies bought by healthcare companies. "That includes products that Owens & Minor doesn't distribute, such as pharmaceuticals, patient food, and linens and scrubs. Also coming soon is a mobile version of Wisdom that will let O&M customers and field representatives access Wisdom reports and information through wireless, hand-held devices" (McGee, 2001, p. 40).

The services do not stop with Wisdom. O&M also offers Web-based order entry and tracking to provide customers with real-time access to pricing information, product availability, and order processing. The firm offers another service that provides alternative scenarios to pinpoint cost savings in a healthcare supply chain, anywhere from ordering to delivery and receiving. The firm exemplifies the concept of using strategy and technology to establish a distinctive advantage in a particular market.

A Business Case Is Needed to Support the Technology

The use of superior technology in a business sense requires a highly developed business case to support that use. Forrester Research has reported that 45% of companies surveyed by that firm are considering CRM projects, from full-blown efforts to pilots. The say a typical firm will spend $15 to $30 million per year on software and services (Leon, 2001, p. 1). That is a serious investment, but our research indicates most of these companies will fail to make sense of their anticipated CRM effort before starting to pay for and play with the technology. Add to that contention the fact that half of the companies implementing CRM have no way to measure the results and you have a formula for disaster.

CRM is all about satisfying customers so you increase revenue. To do that, let us start by understanding that such an effort requires cross-functional collaboration. This is not something for the marketing department. It is not something to be decided in isolation by the CEO. It should begin with an IT executive who is business-wise and knowledgeable in what is available and working. This individual should work in concert with Marketing and Sales to determine how to link the CRM effort with the market segmentation that should have taken place beforehand. The Planning and Manufacturing representatives should have a say in how the effort will either fit operations capability or require more flexibility.

Of critical importance is having someone represent the customer service function, so the representatives, who must show results from using the output of the system, can have a say in what they would like as a deliverable. Giving them a single point of access to most of the information they need in a real-time environment is a valuable asset, but it will only be as good as the enhancement they see to their daily effort. Counsel them up-front so you can deliver what they think has value. This is also a good group to work with — to develop the necessary metrics that have value to the customers.

In the latter area, understanding how technology can increase revenue through improved customer interaction will be difficult to get across to the various functions. You are well advised to have a CFO explain this condition in terms understandable to all of the internal constituents. The bulk of CRM benefits realized so far are still in the area of reducing the cost of supporting customers. Companies are simply not getting to the top-line benefits. One of the most successful efforts I have encountered started with a review from the CFO explaining the potential impact to lines from the top to the bottom of the P&L statement. She quickly received endorsement of the executive management team and a solid strategy ensued.

In short, it takes a team effort with lots of up-front planning to get CRM off on the right foot. That means, before millions are spent on software with spurious intentions, the firm is advised to organize a senior team to develop a business case. Most companies have an enterprise-wide resource planning (ERP) system already in place. ERP contains sales information that is generally not tied to customers at all. There is a good starting point — sharing data that could lead to more sales. Most firms also have done a market and customer segmentation analysis to determine where they should be focusing future efforts. Now you have two ingredients around which to build a business case that results in selling more to the most desirable customer segment. Sharing information of value across the extended enterprise becomes the secret ingredient and the binding adhesive that makes CRM a success.

Next, turn your attention to the issue of control, understanding everyone wants to control the customer. That is probably why CRM is called customer relationship management. But it is not an issue of control. It is an issue of service and attention to the right customers at the right time. Start the focus by determining how to enhance customer relationship building, remembering you cannot manage a relationship that does not exist. Then move to how to segregate the relationship-building effort with a match to what you decided during the customer segmentation effort. Eventually you will find what the leaders have found: you can move CRM into *partner relationship management*, moving down the scale from your most valuable customers to those requiring (and often desiring) less attention and service.

There Are Keys to CRM Success

As the firm builds its CRM strategy, there are a few aspects that appear in all successful efforts. They should be considered as the company moves forward on what will be a very extensive enterprise-wide exercise. My favorite ingredients include:

- *Focusing on business needs rather than technology implementations*
    When a firm introduces CRM without a clear connection to the business needs, it generally falls into a trap. The sales people do not populate the databases with the required customer information and other functions struggle to get data in a form useful for their needs. Consider the problem encountered by tobacco company, Liggett Inc., as reported by Ted Kemp for Internet Week.com:

    Liggett Group Inc. learned first-hand the problems that arise when a CRM application is designed for only one department's needs. The tobacco company gathers spending information and other basic data from stores that sell the Liggett Select, Pyramid and Eve brands of cigarettes through its 60 sales representatives who have frequent contact with the stores. When the CRM project launched, the sales reps used hand-held devices supported by sales force automation (SFA) software from AvantGo to send pricing and sales data back to a central Microsoft SQL database server. Merchandising staffers could examine the data and decide where geographically they should spend promotional dollars. But eight months into the project, merchandisers were still seeing only a dribble of information coming in from sales reps. Salespeople resisted using the application because they got nothing in return, says Mike Lehman, IS manager for Liggett's Western business unit. 'Though it did help out considerably in the office, it really wasn't of all that much benefit in the field. It was just something else for them to do,' Lehman says. To give sales reps incentive to populate the database, Liggett modified the AvantGo application so salespeople could access data that was useful to them, including sales histories and even updates on whether the customer was paying on time. Liggett also added new data fields to the application so employees could enter useful information, such as observations of competing promotions found in stores. Now the application helps the salespeople do their job – selling – rather than merely adding to their responsibilities. Liggett has seen a turnaround in reps' willingness to use the application to manage their customer relationships. (Kemp, 2, 2001b, p. 1–2)

    In plain terms, CRM applications that do not take into account the varied ways different business units, departmental functions, and personnel work together will not succeed. Moreover, the firm has to understand that unless you show the salespeople how the use of technology or software will make them more money, it will not be used.

- *Designing process steps and enabling systems to help the customer, not plugging in software that automates the back office procedures*

    Getting CRM right is all about understanding how every part of a business interacts with its key customers in building a fail-safe system to make it easy for those customers to deal with the firm. This requires new and often unfamiliar approaches, but it certainly means helping the customer so you get more sales in the future. It also means putting a customer-centric strategy ahead of the software purchase. If the objective is to marshal and deploy customer information to enable more effective service and foster more frequent interactions, where appropriate, then the firm is well advised to start with a strategy built around what the customer wants and needs.

    As Elizabeth Herrell, research director at Giga Information Group, puts it, "The key factors for CRM success are designed to reinforce the message that enterprises need to think of customer transactions as episodes in a dynamic, evolving relationship rather than as isolated events" (Herrell, 2001, p. 17). That means linking and reorienting existing processes and channels around the customer. It can require investing in contact centers that provide the kind of after-sale service the customer wants.

- *Aligning CRM objectives with the business strategy*

    Many managers try to ram their CRM efforts through without taking some time up front to win support across the firm, and making sure the objectives mesh with the overall business strategy. In so doing, they fail to evaluate and alter business processes to correspond with the capabilities of the CRM package. CRM is not a product or a single application. It is part of a business strategy.

    Online travel service company Expedia Inc., for example, uses CRM to enhance its strategy of meeting customer needs and building new revenues. The firm uses "customer spending and transaction data to forecast how much it should staff its support centers, how many servers it needs and other customer support considerations. Customer data are also pivotal when Expedia is deciding how to market toward individual consumers. The company captures customer information through inbound e-mail, the Expedia.com site and phone support" (Kemp, 2, 2001, p. 4).

A customer-centric strategy starts with recognizing the value of each customer and being aware of the costs of acquiring new customers. Next, the firm accepts that CRM encompasses the entire firm, and involves integration across disparate functions, from the front-office-facing functions to the back-office inventory and supply chain activities. The linkage has to be of a high caliber, providing valuable data to improve customer relations, through any channel — at the store, by mail, on the Web site, or over the telephone. Mostly, it means offering an improved service proposition and keeping promises to customers. The rewards include increased business volume.

Within a CRM strategy, there are three issues that must be addressed:

- **Organizational issues.** These concern the pervasive nature of CRM and the imperative that every function plays in supporting the customer interaction. That requires a strong manager and leader for the process.
- **Customer awareness issues.** Customers need to know what is available, the value that will derive to them, and how to be comfortable with and make best use of the alternate channels.
- **Technology issues.** Getting the customer perspective right in the first place is crucial, but technology is still vital to success. Integration is the key. All applications must run seamlessly across all channels in the enterprise. Different personnel in different parts of the organization must share a 360° view of the customer being served.

Case Study — Procter & Gamble

How does a consumer products company build a long-term relationship with the largest retailer in the world? How do you enhance that relationship when the retailer has a reputation for the difficulty encountered during the negotiation process? How do you structure your relationship so both firms derive extra benefits, which elude the normal manufacturer-retailer processing? These were the questions faced by Tom Muccio, Vice President of Customer Business Development at Procter & Gamble Worldwide, as he and his team developed a winning CRM program with mega-retailer Wal-Mart.

The enormous success of the partnership between these firms led P&G to launch its customer business development (CBD) strategy, centered on the multifunctional teaming practices developed for this leading-edge alliance. This concept, now deployed globally with nearly 80 teams, is considered a core competency at P&G and defines the working relationship with its most strategic customers. It involved the transition from national to global teams, as well as changes that support multifunctional customer relationship management with the firm's entire customer base.

The relationship being considered was initially fraught with the typical problems confronting such a large base of business — P&G provides more than $5 billion worth of products on an annual basis to Wal-Mart. There were individual ego problems to be overcome, questionable executive support, poor clarity of goals, misaligned objectives, and the usual problem selling the idea across a number of business units and functions. In spite of these complications, a team began, back in 1987, to review the relationship and determine how it could be improved for both parties.

In 1987, the team characterized its relationship with the retailer as one based on a military model with an absolute hierarchy that had to be observed. P&G had a silo approach that dictated sales and went through each product division, no matter how many people called on the retailer. Sales approaches were tactical in nature and interdivision competition was fierce. A new approach to customers was required and two forces emerged from the team's initial assessment. There had to be better methods that would provide better results, a central ingredient of which is a holistic approach with a focus on quality and an end-to-end total system of response.

A telling anecdote described the initial condition. Three years before the team exercise, Wal-Mart officials contacted P&G headquarters in Cincinnati to deliver a "Vendor of the Year" award to P&G. Unfortunately, they were told that the P&G CEO did not take calls from customers. So the award went to another firm. It was not long after that event that P&G took on a greater customer orientation. The driving hypothesis became: If we focused the same "internal" management principles toward the customer, we could expect better alignment, relationship, profit, and sales for both parties. The joint mission statement secured senior management acceptance. It stated:

> The mission of the Wal-Mart/P&G Business Team is to achieve the long-term business objectives of both companies by building a total system partnership that leads our respective companies and industries to better serve our mutual customer — the consumer.

Of interest, when the team mapped the P&G competencies against what the assessment team listed as practices desired by the customer, they found most of the competencies (product data analysis, floor and display planning, etc.) never came into play in the relationship. The best results came from what the team called "little clusters of overlap" where competencies enhanced the overall supply chain process steps — on-time delivery from the supplier, for example, matched with fast turnaround at the retailer's dock.

Next came the operating principles for the emerging relationship:

- Apply performance-based reward and recognition
- Take a positive approach
- Win as a team
- Treat everyone as an individual
- Communicate openly
- Be honest
- Be an owner
- Respect confidentiality

With these very sensible but business-difficult objectives as their guide, the mutual teams set forth to develop the relationship strategy. Consumer research from both firms was shared. Categories were studied to determine where the best results were being achieved. SKU proliferation was addressed to bring the store offerings to those that made sense for both firms. The two companies shared common analytical tools — to diagnose the meanings and trends from their data — and databases so both companies could look at each other's information. Logistics and systems coordination took on high value and activity-based costing and joint business planning went to the heart of the team's structuring of what would be the ultimate CRM process.

To guide the effort, a set of expected results was quickly developed:

- Increase sales, profit, and shares of market
- Reduce costs
- Increase capacity and capability
- Develop precision service
- Introduce speed and innovation
- Demonstrate better consumer insight

Some of the key findings that emerged during the relationship-building exercise have meaning for all firms pursuing CRM. In spite of a high self-perception of its products and the firm's capabilities, the team found customer satisfaction was not what was perceived and was a crucial perquisite to growing share. In spite of offering well-known branded products, some customers were not happy with what they received. Responding to some of those specific problems led to better customer satisfaction ratings. Finance had the biggest problem with the transition. They wanted to measure performance to budget. The best results occurred when the P&G CFO sat down with the customer CEO to discuss how the manufacturer could get proactive to meet the customer's needs.

The team also found the capabilities of account leaders were crucial to the outcome. Constructive, move-forward, effectiveness-oriented, and collaborative attitudes were often not characteristics of the national account sales representative. The sales structure at P&G subsequently went from high silos of product capability to integration within a new format focused on customers. Emphasis is now placed on the "customer experience" as a guide to enhancing the relationship.

The P&G/Wal-Mart relationship has blossomed since this early team work effort. Over 160 multifunctional teams have been involved. Working together on joint goals, a strategic partnership emerged that is essentially unprecedented in the manufacturer-retailer arena. The focus is on the end consumer and what P&G products make the most sense. Shared consumer data lead to focused selling and merchandising. P&G instituted a supplier-managed inventory system by which the racks and floors of Wal-Mart and Sam's Club outlets are replenished by P&G. Data on what to stock comes from daily cash register receipts so replenishment is based on consumption and not sales forecast. It is truly a state-of-the-art relationship that only improves as the teams continue their joint focus on joint opportunities.

Summary

In this chapter, we considered an element of supply chain management that really should be viewed as the price of admission for a company interested in being a decent service organization. Nevertheless, most CRM efforts fail. The reasons have been discussed in detail. The secrets to success begin with an understanding of what the effort is all about and what should be the intended purpose. It progresses with the help of a few willing and trusted customers to develop a sensible business proposition and to design systems that benefit both parties.

At all times, a focus must be kept on the users of the system so they see value for themselves as they transition to what is typically a technology-based information transfer system. Once the concepts are clear, a strategy that fits CRM into the business plan will assure the firm that attention will be given first to what the customer needs and appreciates as differentiating service, and then provide the desired top- and bottom-line improvements. The basics are clear but often overlooked. CRM is a tool to get closer to customers and then build more sales.

7 Mistake 6: Not Focusing on the Consumer

As supply chains become value chain networks and nucleus firms (those with the scale and brand recognition) drive the effort toward optimization, with attention being given to top- and bottom-line results, the involved firms reach a point where an essential transformation must occur. The central focus for the combined effort must move to the end consumers — those people doing the buying that makes the supply chain a viable entity. Satisfying these consumers is the essence of an effective supply chain network and the ultimate reason for creating alliances within the network.

Few networks have this requisite orientation, although all profess to have their eyes solidly on the consumer. (For the purposes of this chapter, customer and consumer will be used interchangeably, although we prefer the term *consumer*.) Our research and that of many other professional firms clearly indicate there is far more lip service given to this focus than actual business practice. Therein lies the next obstacle to effective supply chain management. Any effort is not complete unless it brings a serious focus to the importance of those consumers at the end of the value chain and meets their needs.

Meeting Needs Is More Than Taking Orders and Making Promises

Let me share a supply chain story from the consumer's perspective. My wife and I recently acquired a second home in a warmer climate and set about to purchase furnishings so we could better enjoy the winters away from Chicago. We're multichannel shoppers, meaning we satisfy most of our needs through store purchases, search and satisfy special needs through catalogs, and are hooked on using the Internet for buying some of what we need. We make use of computers, facsimile machines, mail, and cell phones in our quests. As a result of these proclivities, we visited many stores, searched through numerous catalogs, and surfed any number of Web sites where we placed orders for furniture, fixtures, and accessories. In most cases, we were able to get our orders entered efficiently, received confirmation of shipping details, and had the goods arrive at our new domicile in good order. That is the good news.

The bad news occurs at the point where the service we expected ended in an adverse situation — when we dealt with the firms having incomplete supply chain capabilities. We interpreted those situations as ones containing a company with a weak consumer focus. It made no difference to us if it was just one company in the supply chain linkage, we held the nucleus firm responsible. They did not understand our needs, did not appreciate our business, and could not fulfill the promises made through whatever channel we chose for delivery.

Although most pieces arrived in good condition at the designated time and place, a third of the orders went awry. In particular, one attempt to get a special set of dishware resulted in 30% of the goods arriving in pieces, several hopelessly shattered. The order for these goods was placed in a physical store, having what we thought was a good reputation. The promise was that a well-known delivery service would bring the goods to our new address. In spite of packaging that showed meticulous care at the store, some of the containers were so badly damaged in transit, there was no way the fragile contents could have survived the trip.

Several pieces of furniture that appeared to be in stock, based on the firm's inventory reports, were promised within a week, only to arrive months later after we discovered they had to be ordered from the factory. Our conclusion was that the furniture industry, in general, has some missing links in its efficient supply chain processing. They have not completed their homework when it comes to satisfying the consumer. Forget the fancy showrooms and the inventory displays you encounter or the attractive Web sites. The system is only effective if it handles all process steps in the end-to-end connections that lead to my living room (and back to the manufacturer with any returned goods).

Don't pass these incidents off as isolated examples of the problems of within-store, catalog, and Internet buying, the multichannel options available to today's consumers. Supply chains, no matter where they start, end with the consumer. The linked process steps in that connectivity are only as good as the weakest links, and upset consumers will blame the most noteworthy company name. The industry leaders are really distancing themselves from the pack in this area, as they bring elements of consumer satisfaction and e-fulfillment to their supply chain portfolio of skills, through whichever channel is chosen. Generally driven by a nucleus firm, they are focusing their entire supply chain network on specific end consumers and are building a supporting infrastructure that is seamless and effective, from beginning materials to final purchase. The objective is to have a system with no returns and totally satisfied consumers, regardless of the channel chosen for placing the order. That requires a network of allies who see things the way the nucleus firm does.

In this chapter, we will look hard at how this elusive concept can be the secret to building the desired top-line growth. If a firm and its partners can overcome our next supply chain mistake — not paying proper attention to what the consumer wants and expects — they will realize a lift to sales not achieved by lesser networks.

It Starts with Knowing Who Is Doing the Consuming

Most firms have done some sort of customer segmentation analysis, often multiple times, to determine who does the buying and who is represented in the most loyal and profitable segments. They can do so by analyzing sales histories containing consumer demographics or by hiring an outside firm skilled at poring through sales information to develop profiles by whatever segmenting makes sense. The idea is to take a hard look at what specific consumer segments buy, how much they pay, when they purchase, and so forth.

Ito-Yokado, the Japanese Seven-Eleven retailer, has been applying these data for longer than most companies. The firm installed consumer software into its cash registers quite some time ago, and has been building consumer-detailed consumer profiles. Each time a consumer comes into a store, the person running the cash register enters demographic information based on frequent buyer card data, as well as what was included in the sale. Special keys are hit that record such information as estimated age and gender. This company now knows enough information on its best customers that it can send notices to the home advising of special sales on school supplies when a particular family member enters a new level of schooling.

With these types of data, companies can develop strategies on how best to satisfy the various segments they wish to serve. Unfortunately, there are still many companies that cannot explain lucidly who is doing the buying. The information is buried somewhere in their database, but they seem unable to extricate it in a useful manner. That is step one in bringing a consumer focus to the supply chain. As a company moves to the external environment, or the third level of supply chain evolution, it must have an online system that discloses where the current sales are deriving and sufficient data by category to enable sensible decisions regarding further actions. As progress is made with network partners, it becomes imperative that these firms come together to share consumer information and analyze the combined data they possess on who is doing the end buying from their linked system.

Samsonite, the Denver-based luggage manufacturer, has made extensive use of the product registration information provided on the customer notification cards included with its products. It turns out the most loyal customers, with the best data on actual needs, are those who take the time to complete those cards. Now the firm can analyze who is doing the actual purchasing and get a profile because the cards contain a lot of personal data.

Armed with this information, the partners can select obvious end-consumer candidates for focused selling efforts and special promotions, which can be launched and tracked through each of the three channels of contact and delivery. Such efforts can begin in a small way and branch into more elaborate schemes. Office Depot sends selected customers an e-mail newsletter every other week and a promotional offer every other month. Nordstrom.com communicates with consumers weekly, having found that is the correct number for this retailer. The secret for both firms is they now know, from analyzing their data, which consumers should receive the special contacts.

Athletic shoe retailer, Road Runner Sports in San Diego, wanted to build a tight relationship with its online shoppers in a manner similar to what it did for in-store consumers or over the telephone. Moreover, it wanted to offer special treatment to its best consumers, regardless of the channel they chose for the purchase. "So in March, the company began to use personalization to create a tiered Web site that mirrors the lifetime value of each of its customers."

"Using BroadVision's Retail Commerce and One-to-One Publishing software, Road Runner stores use profiles created by their online customers and match them with spending habits. The site automatically shows Road Runner Club members, who typically have a 10 to 15% greater overall lifetime value than nonclub members, more informative articles and special offers than it shows to nonmembers. VIP members, who have a 15 to 20 percent greater value, automatically receive even more in-depth content, including special VIP sale days that offer special discounts to the big spenders" (Colkin, August 27, 2001, p. 49).

It Continues with Knowing What Consumers Want

It does not require a great deal of research to know the retailing world has become a brutal arena. The demise of Montgomery Ward just shows even the large players are vulnerable. As Wal-Mart becomes a mega-giant in this world and specialty stores struggle to sustain their niches, the focus turns (after identifying who is doing the buying) to *what do these buyers want* and *how do I get them to my store*.

Let us look at some basics. According to Texas A&M professor of marketing, Leonard L. Berry, "The key is focusing on the total customer experience. Whether you are running physical stores, a catalog business, an e-commerce site, or a combination of the three, you have to offer customers superior solutions to their needs, treat them with real respect, and connect with them on an emotional basis. You also have to set prices fairly and make it easy for people to find what they need, pay for it, and move on. These pillars sound simple on paper, but they are difficult to implement in the real world. Offering four out of five pillars is not enough — a retailer must offer all of them" (Berry, 2001, pp. 132–133).

Consumers have been analyzed to great extremes by some of the most sophisticated organizations in the world. While there are various opinions on central characteristics, few make sense for firms interested in building a focus on consumers. As consumers consider a purchase, they first want access to information on what they can buy through the channel of choice. That means do not confuse me with fine print and product details too difficult to understand. Do not offer different products and pricing in different channels from the same company.

Next, they want to make the purchase in the most consumer-friendly way possible. That means give me a point of contact where I engage a friendly and knowledgeable sales representative — a very elusive commodity. It is little wonder why so many people today access the Internet to find what they want and then go to a store for the purchase. At least the computer will not be rude to them. And if you think Internet buying is on the decline for the network of which your firm is a part, you had better check the current figures. In spite of the great publicity about dot-com failures, certain areas are still enjoying success with the Web buyer. Keep in mind that appliance retailer Sears has found 10% of all major appliance sales made in its stores are influenced by online research.

Consumers also crave a pleasant buying experience, good service, and easy access to products, again, through the channel of choice. Most retailers believe the most important element is the price of the product.

Research is proving that concept to be a myth. In response to one extensive survey, consumers said "they were less concerned with getting the lowest price than they were with getting a fair and honest price. They want a price that is consistent and that doesn't appear to have been artificially increased or decreased at the expense of other things they want to buy" (Crawford, 2001, p. 23).

In today's hyperactive consumption arena, with retail space probably being twice what is needed to meet demand, selling goods and services cannot succeed without a capability to respond quickly and accurately to what the consumer perceives is needed. They want what they think they need as quickly as possible, but will settle for getting it in good condition on the day promised through the chosen channel of distribution. That requires a supply chain network without weak links. And when you provide those elusive commodities, the results can be an eye opener. "Both Talbots, a Hingham, Massachusetts specialty apparel retailer, and Land's End, Dodgeville, Wisconsin, report that 25 percent of their online shoppers are new customers not on record as having shopped their stores or catalogs in the past" (Power, 2001a, p. 14).

The ability to return goods, through whatever channel was chosen, to the store of choice is another part of doing things right in the eyes of consumers. Multichannel shopping represents an unprecedented opportunity to unlock new sales, but it comes with challenges, not the least of which is the return of purchases. "Customers have told us they want to have a consistent experience across channels," said Dennis Honan, vice president and general manager, Customer Direct, the catalog and online division of Sears. "It wasn't easy to integrate our legacy systems to make sure our pricing was consistent between online and stores, but it was important to the customers, and so the integration became a priority" (Power, 2001a, p. 15).

With the landscape still not clearly defined, a firm may have to feel its way through the maze leading to consumer satisfaction. The best approach is to pilot some applications with specific categories. Consider the experience of outdoor gear retailer, Kent, Washington-based Recreation Equipment, Inc. (REI). "Not only did the retailer find that a traditional store-based event such as a private sale translates well in the online environment, but REI was able to add a special twist not possible in stores. In the brick-and-mortar world, hosting a private sale for VIP customers means closing the doors to other shoppers. Online, however, select customers took part in a virtual private sale while all other online shoppers were still able to shop the site as usual.

"We beat our projection for online sales," said Joan Broughton, REI vice president of online and direct sales. The special promotion involved sending select customers postcards containing a code that gave them access to a portion of the Web site unavailable to other shoppers. During the hours of 6:00 to 9:00 p.m., when stores were hosting private sales of their own, these shoppers were treated to special Web discounts. "In effect," Broughton said, "select customers participated electronically in an event hosted in another channel — stores" (Power, 2001a, p. 16).

Above all else, consumers want credible information about the purchase they want to make. That means going beyond just displaying the products and providing backup information about the goods. It means providing some form of customized full service. Home Depot found its consumers prefer to spend their time working on a home project, rather than learning about the tools they will use or the supplies they need. That is why that firm spends so much time educating their consumers on the project at hand. Walk into any store and you will quickly find the tutorials available on any number of home improvement projects.

And if you think all of this is just about satisfying the retail consumer, you are missing the point. CheMatch.com is a B2B entity focused on multi-million dollar trades between commodity chemical makers and buyers. "In B-to-B, it's not about a pretty picture; it's about getting the junk away from the users and getting them what they want to know," says Roberta Kowalishin, senior VP of business development for the marketplace. "Analysts might want news, but traders moving millions of dollars worth of chemicals don't want news unless its relevant — they want to get to the floor and start trading" (Colkin, August 27, 2001, p. 50).

A Consumer-Focused Strategy Is Critical

Knowing who is buying and what they want, the next question to answer is: How far do we go with this consumer focus? My answer is as far as you can stretch your imagination and resources. The consumer is in charge of where business is going. Satisfying your consumers becomes the newest challenge and the secret to the future. Once your supply chain is in order and you are working effectively with network partners, you must turn your attention and your strategy toward consumers. That means going to the nth degree to satisfy the ones most important to the future of your network.

Consider how far home furnishing retailer IKEA North America went when it decided to carve out a significant share of the furniture and specialties market in the United States. Working with a subset of the company's consumers to develop prototypes of new products, IKEA consumers were asked to dream up their ideal product or service — or to shift themselves into a 'wish mode,' reports Jason Magidson, who helped create the customer design process for the Chicago IKEA store.

The benefits of using this approach are demonstrated in the design of that store. Magidson further relates, "At the time, IKEA was aiming to grow in North America by creating stronger bonds with its customers, in part through the creation of a more compelling shopping experience. So it assembled nine groups of roughly a dozen customers each to get their ideas about the new store's design" (Magidson, 2001, p. 27). To ensure the groups would focus on creating an ideal store with an improved version of the company's existing stores, they were given the following initial instructions: Assume that all IKEA stores were destroyed last night and new ones will be designed from scratch.

Against that backdrop, group members were asked to create a list of specifications for the ideal IKEA shopping experience. Eliciting such ideal specifications requires some guidance and coaxing. Members were encouraged to focus on what they want instead of what they do not want. IKEA's customers were asked to come up with a design for the Chicago store that fulfilled their wish list. To help these customers, IKEA created a three-story octagonal-shaped building with a central atrium that serves as the shopper's home base (Magidson, 2001, p. 27).

What did they come up with? From the central atrium, customers can easily locate the eight departments on each floor. Related products are grouped together; near sofas, for instance, are lamps, pillows, curtains, and CD holders. A restaurant serving Swedish food on the top floor contributes to the store's pleasant ambience. To speed the checkout process, IKEA increased the number of large items shoppers can retrieve from a self-service warehouse. Were the customers satisfied with their design? A survey reported that 85% of people coming to the store rated the shopping experience excellent or very good. None rated it poor or even fair. "Return visits to the Chicago store are higher and shoppers spend an average of one hour longer than they do at other IKEA stores" (Magidson, 2001, p. 28).

From another aspect of going as far as possible, paying salespeople for getting the sale secures their effort and keeps the focus on the consumer. When a new consumer places an order with CDW Computer Center Inc., for example, one of the company's 1200 salespeople will pocket a commission, despite having done no work to win the sale. When the Internet was being interpreted as a system that could displace the Vernon Hills, Illinois-based company's telephone sales representatives, a decision was made to not abandon their call centers. "I don't want one blind order," said John Edwardson, CDW's chief executive. "I want a relationship between the customer and an account manager" (Kaiser, 2001, p. 1).

The journey for this firm has led to a model showing how to merge the Internet with an existing sales force. "Through policies such as giving account managers equal commissions whether orders are won via extensive schmoozing over the phone or just bubble up from the Internet, CDW is saving money without alienating its customers or account managers," says Chicago Tribune staff reporter Rob Kaiser. "Today, the company reports that 15 percent of its revenue comes from direct Web sales, where customers themselves click and key orders into CDW's site," Kaiser adds. "The appeal of such sales is obvious: Companies spend an average of $30 completing a phone sale, but that figure shrinks to $1 or $2 for Internet transactions" (Kaiser, 2001, p. 3).

CDW has developed what could be considered a unique Web model, using Internet transactions with phone support. Following an emerging pattern in which Web surfing and personal contact are combined before the final sale, the CDW account managers often consult with customers before orders are placed online and follow up after the sales are finished. The sales people also keep track of accounts as they approve all orders, even those arriving online. "If a company that always gets Toshiba laptops suddenly orders IBM Think-Pads, the salesperson will call the company to inquire if an error was made or if a new project is underway, which eventually could yield more sales. Having more customers entering orders themselves allows the account managers to dig deeper into what technology products their clients will need" (Kaiser, 2001, p. 4).

In a further step to maintain an element of customer intimacy, when a consumer signs on, a picture of the assigned CDW sales person appears on the screen, as well as whether the person is in the office or temporarily out. The account managers' identification badges trigger these messages as they enter or leave their offices.

Having Multiple Channel Capability Is Essential

One thing we must keep in mind as we consider catering to the needs of the current consumer is that it is a multichannel environment. You cannot just offer me one way to make my purchase and expect me to view you as a modern establishment. If not for me, for someone in my family, you must offer options. And there is money to be made in the process. "As shoppers' service expectations continue to rise," says industry analyst Denise Power, "many multi-channel retailers are asking, 'How high?' Today's multi-channel leaders are happy to respond to these heightened demands because they've seen revenues and customer bases grow when shoppers are given what they want when they want it in the channel they choose" (Power, 2001a, p. 14).

Power cites The Limited, a Columbus, Ohio-based retailer, as a firm that offers its consumers buying opportunities in all three channels — store, catalog, and Internet. It is just good business. "Office supplies retailer, Staples, Farmingham, Massachusetts, estimates its multi-channel shoppers spend up to four times more than single-channel shoppers."

Consider another interesting trend that meets the needs of the multichannel shopper — the installation of in-store kiosks. A personal computer hooked to BlueLight.com in Kmart stores has turned into a new means to generate sales. According to Ellen Neuborne, writing for *Business Week*:

> Kiosks, once regarded as a low-rent customer service desk, are rising to a new role: pitchman-in-chief for the company Web site. Retailers have spent big on e-commerce, but they have no way to direct shoppers to their virtual stores. TV ads cost too much, banner ads don't bear much fruit, and you can generate only so much buzz plastering a Web address on bags. Kiosks bridge the gap between brick and click. (Neuborne, 2001, p. EB 6)

Retailers are apparently beginning to see a positive impact. Five months after installing 3,500 kiosks nationwide, Kmart reported that 20% of shoppers at BlueLight.com came from inside Kmart stores. That makes the kiosk buyers the second strongest draw for the e-tailing site, behind only Kmart's free Internet service. Evidence shows shoppers who purchase both offline and online from a given retailer tend to buy more. Neuborne also cites Recreation Equipment Inc., which added an e-tailing arm and kiosks in its 60 stores. "Customers who shop both online and in stores spend 22 percent more than those who buy only from the real-world outlets," says Joan Broughton, REI's vice president of direct an online sales (Neuborne, 2001, p. EB6).

Without a doubt, e-retailing is a reality and a portion of sales is going to move through this channel. And what does it take to succeed in this area? According to John Stuart, U.K. director of global e-business solutions company Merant, "In a nutshell, there are three areas that can be considered critical success factors — choice, personalization, and service quality. If there's no choice — or even not enough choice — it will lead the visitor to call on other sites. Personalization will build loyalty and encourage sales — and service quality means that the fulfillment part of the process is painless for the customer" (Stuart, 2001, p. 47). These comments apply to buying in general and set the stage for the next requirement, covering the last mile in the delivery process effectively.

E-Fulfillment Must Be a Part of Advanced Supply Chains

Whether a consumer makes a purchase in a physical store or places an order over the Internet, which we project will continue to increase as a percentage of new sales but will be limited by industry to a small percentage of total purchases, the goods have to get to the point of receipt. Success in any retailing operation, or any good business delivery system for that matter, depends on whether or not the firm can fulfill orders in a manner that meets consumer expectations. Rather than simply focusing on getting orders out of the factory or distribution center so revenues can be booked, the astute company concentrates on doing the job right the first time and retaining consumers because of the skill in their delivery processing. That is becoming the heart of the second wave of fulfillment, often enhanced through electronic capabilities built into the processing. Moving to the most advanced levels of supply chain requires a firm to enter the realm of electronically enhanced fulfillment —or e-fulfillment.

Patrick S. Sedlack, writing for *Supply Chain Management Review*, puts the necessary new focus and transition into proper perspective, using e-tailing as the medium of distribution.

Moving from the first to the second wave of e-fulfillment begins with an understanding of the difference between e-fulfillment and traditional fulfillment. Online orders typically are picked and packed in 'eaches' and then moved via parcel carrier to the customer, who is usually located at a residence. By contrast, traditional shipments move as palletized freight in truckload quantities, typically replenishing a store or distribution center. E-Fulfillment also differs significantly from catalog fulfillment in spite of obvious similarities. Some of the more prominent differences include smaller orders, increased returns, higher customer expectations, and unpredictable demand.

To succeed in the online world, bricks-and-mortar companies that are skilled at pallet load and caseload operations must learn how to handle high volumes of 'eaches' — orders for individual items. Transportation processes that develop truckload and less-than-truckload (LTL) shipments must be adapted to handle online orders for one or two items to be shipped in packages. And companies that traditionally have experienced low levels of returned goods now must cope with return rates of 10 or 20 percent — or higher (Sedlack, 2001, p. 83).

Today's consumer demands near perfection from the network of choice. They want to receive exactly what they ordered in excellent condition at the time promised by the person making the sale. It all sounds easy, but it is straining some networks to the point of driving them out of business.

One of the first requirements is to have a convenient means of accessing the flow of goods across whatever channel is used at all times of the day or week. Consumers expect easy contact with the person making the sale and expect assistance in tracking deliveries once the order is placed. They also expect some form of customization to their orders, even if that includes only a special gift card or greeting placed inside the delivery package. All of this means a firm must have its information online as it relates to availability of goods before order placement, movement of the goods in transit, and a follow up on how well the goods were received. Consumers abhor having to back order goods that they thought were in stock and will switch to another source as fast as they can drive to the next store or move their mouse to a new site.

Attaining excellence in this demanding fulfillment environment begins internally. At the heart of an efficient e-fulfillment system is a sound, integrated pick, pack, protect, and ship operation. Current technology systems should be in play to enhance the effectiveness and speed of the processing. The physical center used for fulfillment must be segregated to handle the variation in order size, from large corporate accounts and full truckloads to individual consumer orders shipped in lot sizes of one to a few. The order management system should then have capability across any mode chosen for delivery — truck, van, train, airplane, or ship. And today, no system is complete if it cannot handle global shipments.

Case Study — Gallery Furniture

Jim McIngvale, President and CEO of Gallery Furniture in Houston, Texas, describes the ultimate example of attention to the consumer. McIngvale opened his first store in 1981 with $5,000 in cash and a viewpoint about selling that would last his lifetime — a commitment to boundary-less customer service. He now carries out that philosophy in a single, 100,000-square-foot store, through which he sells $17 million annually with inventory turns of 35.

When asked to explain what he thought was the key to success in the retail furniture business, he responded crisply, "It's all a matter of paying attention to diversity and matching sales representatives with consumer profiles." His elaboration fleshed out these ideas: his customers represent a diverse construct, including Caucasians, Hispanics, African Americans, Asians, Gays, and Lesbians; carrying out his concept requires him to have a sales force that mirrors those backgrounds; he even has sales reps capable of dealing through sign language. In his store, the consumer gets face-to-face with someone that speaks the appropriate language and understands the person's feelings.

He has spent $400,000 on a playground for children when most competitors display signs advising consumers to control their children or do not allow them in the stores. He spends $25,000 per month on pies and cookies for these kids and their parents. Ten cashiers speed these consumers through the checkout stations. A few others take care of the accounting and financial matters, while the bulk of the balance of the 300 employees are charged with helping people get what they want right away. Same-day delivery is a standard for Gallery, even when it means moving a 60-inch television into the second story of an old Houston home. That one required several trucks, moving the product across a roof, punching a hole in the wall of the house, and cleaning up the whole mess once the set was in and running.

"If your customers are not the alpha and omega of your business," McIngvale advises, "you'll go broke." And he practices what he preaches. His biggest sales time is the 7 days after Christmas. During that time, 200 employees staff the floor so they maximize "face time with the customers."

There are numerous stories of how Gallery goes beyond the Nordstrom-type service model, but one in particular stands out. On June 8, 2000, a couple purchased $18,000 of furniture near the closing time of 9:30 p.m. on a very rainy evening. End of delivery is typically 8:30 p.m., but the couple wanted delivery that night. Gallery employees pulled a special truck to accommodate the delivery, only to discover that about 3 feet of water had accumulated on the route to the couple's home. It became an evening few Houstonians will forget as the city became essentially impassable and water damage was horrific.

McIngvale and several employees stayed all night to raise the goods above the rising water, only to witness enormous damage to their inventory. "Mattresses were literally floating around the building," he recalls. By 4:00 a.m., the water had receded and, by 4:30, the firm had incurred $20 million of damage that was not covered by flood insurance. After considering his own plight, McIngvale's thoughts went to his consumers. He thought that they would probably need bedding. He placed an immediate call to supplier Serta and ordered as many mattresses as could be delivered to his location. Serta agreed to supply the bedding at cost and Gallery announced it would pass on the goods at the same cost; 17,000 sets of bedding went to customers needing a dry place to sleep.

Gallery represents a firm in an industry often marked by poor service, from the consumer's perspective. At this Houston store, a diverse and dedicated sales force makes every attempt to match consumers with similar backgrounds, pay attention to the details of the purchase, and go the extra mile to make sure everyone is satisfied with the experience. Gallery stands as a model of a firm dedicated to its consumers.

Summary

As a supply chain improvement effort matures, it reaches the point where the focus must move to the end consumer. Most firms and their allies profess a strong orientation toward these consumers, but many firms — especially those upstream in the supply chain — have little contact or direct interest in consumers. True business leaders dismiss that thinking and are focusing their networks on these end buyers while building an infrastructure that satisfies their needs through multiple channels.

The secrets include finding out who is doing the buying and then determining what they want in the processing. That generally includes more than just a low price. It means having a supply chain infrastructure that is consumer focused, from end-to-end. With this information, a nucleus firm and its partners can segment the consumers by need and match the service they expect through a seamless network of response.

8 Mistake 7: Misunderstanding the Internet

As network members begin to cooperate and determine they will move their supply chain effort toward multichannel responses for targeted consumers, they enter levels 4 and 5 of the supply chain evolution. Now we are looking at how to create full connectivity across the value chain network and other problems appear. Foremost among those difficulties are understanding what the Internet and Web-based technologies are all about, what the ramifications will be for business practices, and how can these new tools be leveraged for benefit. There has been far more mystery than knowledge in these areas, and that condition has stalled progress for many firms.

In this chapter, the focus will be placed on why the Internet has had such a roller coaster ride when it offers so much promise in terms of improving the processing that occurs across an efficient flow of products, information, and finances. We will cut through the hype surrounding the dot-com crash and consider how the real cyber revolution is about changing the way business should operate and how a few of the industry leaders are tapping into this new technology to enhance business performance. And we will draw some logical conclusions based on what the leaders are accomplishing, while taking advantage of what will be one of the most important changes to affect business processing. Mostly, we will clear up the confusion behind the next supply chain mistake — misunderstanding the Internet.

We Are in Wave Two of the Cyber Revolution

We are slightly beyond the midpoint of a global, cyber-based revolution, which will involve three waves of attack before it reaches equilibrium. The first wave began some time in early 1995 and lasted for more than 5 years. A few visionaries and many would-be entrepreneurs hoping to get rich launched that wave. The idea caught on like wildfire and we were off on a mad roller-coaster ride without knowing where we were headed. Just announcing that you had something that looked like high technology and was related to a marketplace gave you a chance to cash in on the hysterical rush to buy into Internet-based propositions.

The overwhelming winners of the first wave, those smart enough to cash out before its conclusion, became a new, wealthy, and powerful elite. The secondary winners were the consumers in the B2C portion of the cyber wars, who found a new sales channel that increased their control over decisions regarding where to make their purchases — multichannel buying was introduced. A third group, clever purchasing people who discovered amid the hype there was a way to find better pricing for nondirect materials and supplies, learned how to use electronic auctions. Mostly, there were nonwinners, people who for a moment made a killing, only to see the paper benefits drift into the wind. And in the business world, there were people who didn't have a clue what was happening.

Exhibit 8.1 summarizes the three waves we are considering. The first wave was certainly focused on the possibility of achieving riches. Unfortunately, most of the endeavors marking this wave were based on weak business models that made no demand for a profit. That was to come later, we were

**Exhibit 8.1
The Cyber Revolution**

- Wave 1 — 5 years — Hype, hysteria, riches, poor understanding of the cyber impact, and weak business models based on using a new media of communication, with or without a profit, to secure customers and consumers
- Wave 2 — 3 to 5 years — Professional approach with greater focus on B2B, need to generate a profit, and use of collaborative commerce as the major implementation tool
- Wave 3 — 2 to 3 years — Mergers and acquisitions that will lead to a few cyber oligopolies dominating industries

told, after the customers and consumers were hooked on using the Internet at particular sites. The people behind this wave could not wait for their Web site to come up or for the expected IPO that would launch their wealth. The poster children of this wave included Amazon.com, Dell Computer, Cisco Systems, AOL, and Yahoo. There were others, for a moment, but their business models and value propositions were flawed and they were relegated to the trash bin when the first wave ended. Exhibit 8.2 illustrates one well-known example of the rise and fall of what appeared to be a sure-fire Wave 1 investment.

For a time it worked, but by the holiday season of 1999, we had an indication there was trouble. While cyber-based buying was climbing at a geometric rate and we were losing track of the endless trading exchanges, net marketplaces and dot-com entities were appearing, gifts were not showing up, and problems were appearing faster than orders could be entered. Most of the cyber entries forgot they needed a solid supply chain to support the ordering system they had created. Buyers also became a bit psyched about giving away their credit card information to unknown entities. A few observers could see that the inevitable slide down the mountain was going to be as dramatic as the swift climb upward.

When January 2000 arrived and attention could be taken away from the Y2K crisis, the focus moved to the lack of sensible business propositions

Exhibit 8.2
Cyber Revolution — Wave 1 Example

- Chemdex announces a marketplace for life science, biotech products — 1998
- Proposition: $9.5 billion online potential, 250K products, 120 suppliers
- 1999 results: $1.5 MM sales ($305K loss)
- 7/27/99 IPO: $880 MM 1st-day market cap
- February 2000: market cap reaches $5 billion (for a venture selling $1.5 MM)
- March 2000: name changes to Ventro; market cap reaches $8 billion — end of Wave 1 is in sight
- September 2000: stock has fallen from $248/share to $20; market cap declines from $8 B to $660 MM — welcome to Wave 2
- February 2001: Ventro shuts down Chemdex

behind most of the new cyber entities, and the slide began in earnest. The unofficial end to Wave 1 came on April 14, 2000, when the technology sector crashed and a trillion dollars of value was eliminated from the overpriced stocks representing many of the Wave 1 leaders in 6 hours of trading. By the end of that year, about $3 trillion in wealth had vaporized from the wild ride.

Professional Management Will Characterize the Second Wave

The second wave has already started and will last 3 to 5 years. It is being characterized by the entry of professional management, with a greater focus on B2B applications. With better business sense and a desire to take advantage of the Wave 1 learning, these pros are moving into those firms seeking a leadership position in their industry. It is the period when large, established firms that have been observing the cyber phenomenon from the sidelines will emerge to the forefront. They will introduce the e-business modeling that will define the real advantages of Internet-based communication systems.

There Were Those Who Understood and Those Who Wanted the Old Way

Two primary propositions were at work in the first wave of the cyber revolution: use of the Internet to secure a consumer base at all costs and to get rich in the process. While a few well-intentioned technology pioneers were introducing a new media of communication, a horde of new entrepreneurs decided to cash in on the results. The trick was to hook consumers on using a particular Web site while the entrepreneurs appeared as new intermediaries in typical supply chains. In the absence of any real understanding of business values or sensible business propositions, these early cyber players launched into a drive to secure a significant base of loyal consumers, while totally neglecting the need to make a profit.

Investors could not wait to give these entrepreneurs money. In fact, so much money flowed that the whole balance of capitalization was upset for a time. Noticeably missing from the propositions being foisted during this time was a discipline that required a return on the investments being made. Some of the new players burned the influx of cash so fast that they were penniless by the end of the first wave. A legacy of Wave 1 is that there were far too many entrepreneurs and a paucity of business sense. The revolutionary battlefield became cluttered, understanding of sound application logic was obscure, and the need to promote a soaring stock price prevailed. The cyber battlefield is now littered with the remains of those entrepreneurs who did not have the business savvy to back up their grand ambitions. A typical supply chain lament was that they could not fulfill the orders received.

On the sidelines was another group of somewhat older and wiser business people, many longing for a more stable situation in which tried-and-true practices would prevail. This group was often afraid to make a move into the cyber arena, but did so at the urging of aides determined not to be left out of the action. This group included old economy hard-liners who delighted in the crash of many of the cyber darlings. They reminded those who would listen of predictions that e-business was not a solid concept. Most were very happy to settle back into their overstuffed chairs and discuss business in the usual manner as Wave 1 ended. There is a problem with this attitude. This group is unprepared to take advantage of the changes occurring in Wave 2. The most important learning from the first wave is that winning the revolution will require strategies and generals missing in action in Wave 1. The Internet is not going away; it will be a part of future business strategies and success. Those who do not understand this proposition need to increase their awareness — quickly.

Technology is an inevitable process. We have known that for some time. But since the time of the Luddites, there have been those who would try to eliminate the progress it can bring. As the cyber revolution took us on a merry ride, there were those who knew it was overhyped but heading to a different and better way to conduct business. There were also those who were not sure. Some took the time to learn what the revolution was all about, while others remained in the dark. After years of studying supply chain and the emerging need to enable advanced efforts with technology and collaboration, I have one major conclusion: some business managers still do not get it. Call it reluctant conventionalism; there are industry leaders who simply do not want to change the traditional way of doing business.

Louise Kehoe, writing for the *Financial Times,* addressed this issue when she said, "There is a crisis brewing in the corner office. If truth be told, a lot of chief executives are information technology illiterate and the smart ones are uncomfortable with their lack of knowledge. As IT consumes an ever-larger proportion of capital spending budgets, too often the person signing the cheques has only a vague idea of what he or she is paying for. Worse, when it comes to strategic planning, the chief executive who is not tech savvy is at a severe disadvantage. The issue is not whether the chief executive is proficient in the use of a personal computer, but whether he or she can bring a general understanding of IT issues to business decisions" (Kehoe, 2001, p. 1).

These managers are missing the boat. Internet technology can bring new efficiency to the business networks that back any supply chain, leading to better customer satisfaction. Indeed, most analysts now confidently predict the greatest benefits from Internet technology will accrue to the B2B portion of the value chains, emerging as the logical evolution of advanced supply chain and e-business applications. As business and technology practices continue to converge, the drive for riches and a solid consumer base will have had a beneficial effect, teaching that a more sensible approach to applying one of the most significant events of our lifetime can lead to a superior business model.

As Thomas Foster, editor of *Supply Chain e-Business* magazine puts it, "The real e-business revolution is about changing the way businesses operate. It's about companies working more openly with outside partners to eliminate wasted time, resources, and duplicative effort. It's about shared, streamlined business processes, building new business models with inexpensive virtual assets, and laser-like focus on truly strategic activities" (Foster, 2001, p. 8). To fully benefit in the second wave, the executives being mentioned must avail themselves of the learning so they can better lead their organizations through this second wave of the battle. IT executives, for example, are well advised to set up tutorials for these individuals.

But making progress in this area starts with finding out what you do not know. A study conducted by the Economist Intelligence Unit (EIU) and Meritus Consulting recently confirmed the issue being considered:

> In responding to the survey, 82 percent of senior executives said they believed Internet technology would have a major impact on — or even totally transform — their supply chain performance within the next three years. No big surprise there. Yet the vast majority of those same companies acknowledged that they were ill prepared to integrate this technology into their business processes and supporting systems as they currently exist. Notably, only a meager four percent of respondents believed that their customers were satisfied with their supply chain performance. (Ljungdahl, 2000, p. 82)

Technology and collaboration are the tools of this new revolution, but at the heart of what is going on is dealing with global competition, more demanding customers and consumers, and a rethinking of how firms can come together and use the tools to increase top- and bottom-line performance.

Those who fail to see this risk being caught up in the same kind of hype and the problems with the technical sector as occurred when the dot-com frenzy was at its height.

Internet Use Will Generate Significant Business Value

If there has been one lesson to be applied from evaluating the first wave of the cyber revolution, it is that avoiding exploitation of the Internet can put a firm at a disadvantage in its marketplace, while indiscriminate investments and applications will be costly efforts with little received value. Make no mistake — the revolution is not over. There is still time to dominate an industry through reasonable application of cyber technology. The real benefits are coming from seamlessly integrating data flows and work processes across extended enterprises. The desired improvements will emerge when industry leaders understand how to generate business value through such integration. The revolution continues to rage at the convergence of sensible business plans and application of enabling technology. The emphasis must be on value propositions that make good business sense and on issuing reliable promises regarding what can be achieved.

Single, vertical firms cannot win the battle alone. They need the help of their value chain partners. Those firms that make the right alliances now and deploy the proper resources will win the second and third waves of the revolution. This wave will be characterized by an emergence of some very traditional firms in collaboration with some of the dot-com survivors. Consider that Amazon.com has signed a deal with Circuit City in which Amazon will sell the retailers electronics online and let shoppers pick up their items at the retailer's stores.

A Framework Is Needed to Complete Wave 2

A firm making its way through the second wave must answer some important questions. First, *with which other firms should it ally its resources, given that no single firm is going to dominate an industry?* In the future, the competition will be between competing value chain networks and weak links will be fatal to the best of business organizations. Selecting the right business partners is a crucial event to be completed in Wave 2.

Consider the situation emerging in retail sales. To gain leverage by aggregating individual buying power, Auchon, Best Buy, CVS, JCPenney, Kmart, Walgreens, and others have joined forces in the Worldwide Retail Exchange.

Not to be left out, Sears and Carrefour, the French retailing giant, have formed the Global Net Exchange. Meanwhile, the largest of the giants, Wal-Mart, has introduced its online private trading hub. Clearly, the competition is lining up with network alliances as a new order of differentiation. There will be suppliers participating in multiple networks, but they will be working to help that network gain coveted new sales. This trend will continue as the automotive, aerospace, banking, consumer goods, healthcare, high technology, and other industries sort through the options and enter into network alliances they believe will enhance their future prospects.

What we are witnessing is the appearance of spheres of confluence. Partners and exchanges are still working on value propositions and the myriad of details required to establish such entities. But the desire to use the Internet and associated technology and software to gain the next level of advantage is implicit in the actions. Old enemies are expected to collaborate to make the savings. At the center of this development will be the use of the newest business tool — collaborative commerce — the logical extension for sharing the best practices emerging from the first wave via the Internet.

The second question goes to the heart of what faces firms in the second wave of cyber activity. Firms must determine *what the impact will be to their business from the unstoppable attention being given to e-commerce and supply chain.* They will have to sort through the remaining hype and the reality of what can be done to enhance what is fast transitioning into the value chain of connectivity from primary supply to consumption. They will need to decide what percentage of future business will be Web-based by the year 2004 or 2010, as well as how ready they are for the transition.

The third question concerns *what percentage of future revenues and profits will be at risk if a firm does not project a cyber capability.* This requires knowing industry trends, what firms will demand Web access to data, what market segments will be dominated by Web activity, and more. Established companies must be alert to the nontraditional customers they could secure via the Internet as well as those they could lose by not having such a presence.

The fourth issue relates to the fact that, because all channels of distribution require both physical assets (trucks, warehouses, packaging equipment) and cyber capability (order placement, inventory visibility, order tracking), both areas need to be technically enhanced. No firm can perform all process steps; therefore, each must decide *what new partners will be needed to complete an e-business model.* As mentioned, we might select suppliers servicing other networks, but for our effort, they must help us bring enhancing values to differentiate us in the marketplace.

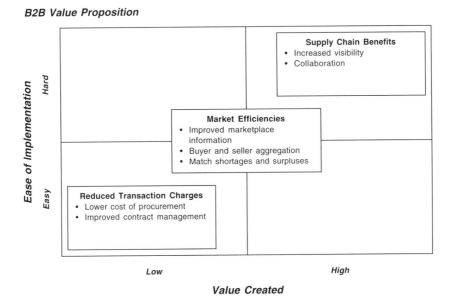

Exhibit 8.3 Wave 2 framework. (Source: Chopra, Sunil et al., Northwestern University, Evanston, IL. With permission.)

The issues described above are not easy to address, but delaying too long to find the right answers can exclude a firm from the best networks and place it at a disadvantage. Getting to solid answers also leads to the creation of a framework for competing in the brave new digital world. Sunil Chopra et al. have provided such a framework, illustrated in Exhibit 8.3. (Source: Chopra et al., Northwestern University.)

According to Chopra, "The Internet's unique characteristics will allow businesses to create significant value in the future. The value of B2B e-commerce, however, will vary depending upon the supply chain strategy and competitive environment that a company faces. Successful companies will be those that can tailor their e-commerce initiatives to support those areas where maximum value can be extracted. Three distinct categories emerge where B2B e-commerce can be applied to extract value. These are reduced transaction charges, improved market efficiencies, and enhanced supply chain benefits" (Chopra, 2001, pp. 50 and 51).

Using ease of implementation and value created as the two dimensions for his model, Chopra positions the three categories between easy to implement and low value created, to hard to implement and high value created, respectively. Beginning in the lower left quadrant, transaction costs are those associated with completing a buy or sale. They include handling proposals and quotations, processing orders, staffing the buying function, operating call centers, and so forth. Using telephone and fax access requires more people for both the buyer and seller, and they contain historically high error rates. As firms introduce a fail-safe, electronic system, the errors and reconciliation go away, fewer hands are needed to process orders, and the process takes less time. Savings from these improvements are the first level in the value proposition.

Chopra elaborates, "The magnitude of the savings will vary depending on each company's specific situation. i2Technologies estimates that companies can achieve transaction savings of close to two percent of sales by using the Internet. Eastman Chemical estimates transactional savings of close to four percent of sales, while British Telecom claims to have reduced transaction costs associated with procurement by ninety percent using e-commerce" (Chopra, 2001, p. 53). Certainly, a company can expect to reduce the cost of order entry from $25 to $100 to something less than a dime. They will also save almost all of the time and cost now going into reconciliation.

Market efficiencies, near the center of the model, offer several areas of opportunity. Improved marketplace information can be leveraged to move closer to optimized conditions as buyer and sellers find what is really available in a particular market. The Internet offers a unique opportunity to aggregate what is being sought and what is being offered. Through this medium of communication, there is an ability to match surplus capacity in the supply chain with demand. Companies in the automotive and heavy-machinery industries have reported savings from 2 to 20% through use of auctions, for example. Per Chopra, "The value of matching surplus supply and unmet demand is likely to be the greatest in industries that experience highly uncertain demand and where flexible supply can be diverted. General Mills saved seven percent of its transportation costs by implementing a backhaul exchange with its business partners" (Chopra, 2001, p. 53).

Supply chain benefits, in the upper right hand quadrant, represent the highest order of potential. Here we find activities relating to the increased visibility of critical information and the opportunity to collaborate electronically with valued partners. Chopra elaborates:

Collaboration is the ability of different stages of a supply chain to use the common information obtained from visibility to make decisions on product design and introduction, pricing, production, and distribution that will allow all partners to profit. For example, Wal-Mart and P&G increase visibility when Wal-Mart shares point-of-sales data. The partners only realize full value, however, when they use this information, along with capacity information at P&G, to decide the best timing for promotions and resulting production plans. If decisions are made independently, Wal-Mart may run the promotion at a time when production costs for P&G are high. Through collaboration, constraints on both sides are considered in determining a schedule that maximizes profits (Chopra, 2001, p. 52). Exhibit 2.1 in Chapter 2 compiles a full range of the potential savings from supply chain activities.

Internet Strategies Will Drive Greater Accomplishments

Use of the Internet is still coming into its own as a value-adding tool for business. In addition to the areas listed in the framework, it can have significant impact on other traditional business processes. Larry Carter, CFO of Cisco Systems, explains how his firm decided to "radically accelerate our financial reporting process." At the beginning of their effort, it took Cisco 14 days to close their books, and there was an urgency for this fast-growing company to get their hands on more current information. The finance department decided to set some aggressive goals. Per Carter it was decided, "We would generate consolidated financial statements in one day, cut finance costs in half, and transform the way we supported the company's decision makers" (Carter, 2001, p. 22). The firm exceeded those goals as it has achieved what they call the "virtual close."

It was a 5-year effort, but what took place was worth the effort. Cisco first established quality standards — and metrics — for all of its data-collection activities. The firm standardized the definition of bookings and backlogs, consolidated responsibilities for accounts payable and purchasing, and eliminated practices that yielded little financial gain. By 1998, "we had globally consistent information available online for Cisco's decision makers," Carter noted. Today, Cisco updates its bookings, revenues, and product margins by the minute. The firm can literally close its books within hours, producing financial statements on the first workday following the end of any reporting period. Cisco also shares financial data with outside suppliers. Daily information about product backlog, product margins, and lead times triggers decisions throughout the supply chain. Suppliers use the information to order inventory, adjust manufacturing capacity, and anticipate shipment volumes. Carter reports, "They can, for example, review sales forecasts for a particular product and plan their inventory accordingly" (Carter, 2001, p. 23).

In a move to make transaction savings, transportation company J.B. Hunt has introduced an integration project to move its biggest customers off expensive EDI networks and onto new Web-based trading networks. The firm started with its 12 highest-dollar customers, which represent 30% of its revenue. Providing the Internet-based solution allows J.B. Hunt and its customers to "move three main transaction sets, which include load tenders (the electronic details of a new job), invoice and payment information, onto the Web network."

Lana Magnold, application systems manager, notes, "J.B. Hunt and its customers will save about $12,000 per week just on that one application. The new project will reduce the programming necessary to bring up new integrations from 40 hours to about 8. That will let developers focus on new e-business applications, particularly self-service interfaces, that will drive new business in tight times" (Karpinski, August 15, 2001, pp. 1–3).

These and other moves are part of a general effort to build Internet technology and cyber strategies into future business models and plans. The roadmap is now available. As executive learning increases and the action studies continue to come in, there is no choice but to get back on the coaster for another, more sensible ride. This time the ride will be made with a few trusted allies and have a defined purpose for the effort.

Extended Enterprise Collaboration Is the Next Frontier

When considering how to make the most of an extended enterprise effort, the key question is not whether the members of the network should deploy Internet technology — companies have no choice if the want to stay competitive — but how to deploy it.

We submit that getting better as partners across the extended enterprise must be included in any advanced supply chain strategy. Internet technology is making companies redefine their business propositions as they seek to improve extended enterprise processing. If we call this effort e-business, then we can say that new e-business models are required for any firm in the B2B2C end-to-end value chain. Players anywhere in that linkage are advised to consider the influence of the Internet and make use of Web technology a part of their future planning.

The Internet and the emerging business models should motivate significant change in supply chain design. That is done best with the help of willing and trusted partners, working for mutual benefit. Such an effort requires firms to transcend normal business models and transform relationships with suppliers, distributors, customers, and now the end consumers. The objective has to be to apply the new technology in a way that responsiveness and efficiency are achieved.

Take a look at what is happening at General Motors. With sales of cars and trucks sliding downward, the company is turning to the Internet for help. Its e-commerce unit, e-GM, "generates 1,000 sales leads per week for its dealers and incorporates Web technology in its cars. Its president, Mark T. Hogan, says the corporate parent remains a big believer in the Internet and how it will reshape the way it will do business." Jeffrey K. Skilling, CEO of energy supplier Enron, echoes this sentiment as he predicts, "Incumbent companies have to come to grips with this new technology because it's very, very powerful" (Hamm, 2001, p. 126).

Business Week reporter Steve Hamm and colleagues advise that, "What Skilling and others are doing is integrating the Internet into every nook and cranny of their businesses. Call it managing by Web. They are using the Net for everything from filing expense reports and calculating daily sales tallies to sharing employees' intellectual capital and communicating instantaneously with suppliers." As an example, he cites, "About 40 percent of Dow Chemical Co.'s customers reorder chemicals without human intervention when sensors in their storage tanks signal they're running dry" (Hamm. 2001, pp. 126–127).

Case Study — Intel

Andrea Williamson, e-business group marketing manager for Intel, explains how manufacturers and their customers can reap rewards as they move forward together using the Internet. On July 1, 1998, Intel took its first order over the Web. Today, the company averages $2 billion in monthly e-commerce revenues and is moving rapidly toward machine-to-machine automation. The initial scope of the effort was to provide the infrastructure, tools, and communications technology for Intel's product groups and sales force to do business securely, efficiently, and effectively with OEM and distribution customers worldwide. The company has come a long way from that position as it discovered the hidden values in Internet technology and collaboration.

By the end of 1998, new systems enabled approximately 200 of Intel's OEM and distribution customers in nearly 30 countries to place orders for products, check product availability and inventory status, receive marketing and sales information, and receive customer support, all in real time. By July 2000, almost 1,000 individual customers in over 60 countries used over 18,000 discrete Intel e-business sites.

Williamson explains that, "Intel's original goal was to move $1 billion of current business away from the fax and phone and onto the Internet by the end of 1998." The company achieved that goal in only 15 days and then quickly ramped to taking $1 billion in orders per month through its e-business system. By 2000, they were taking 95% of orders electronically with 95% of customers. Intel conducts nearly 100% of its business over the Web in Latin America, Taiwan, and Japan.

The journey began in 1998 when a small team of software engineers focused on developing an Internet Order and Inventory Management application and integrating it into Intel's existing finance and ERP systems. The team created a suite of Web Order Management (WOM) applications, a real-time interface with the ERP back-end for customers. At the time of its rollout, customers could do six activities:

1. Check pricing and availability
2. Place orders
3. Check status of backlog requests
4. Enter Hot List requests
5. Check order status
6. Request change orders

Based on customer feedback, the company continued to improve the six features. Intel's 1999 mission was to use the power of the Internet to create significant value for the company, its brands, and customers. The specific intent focused on enhancing customer relationships, bringing products to market faster, and improving overall efficiency. At the heart of this effort is a new, user-centered design based on a massive effort toward more workflow analysis and data collection with customers. Intel and its customers designed the business functionality for the next application. Intel has also offered personalized document delivery to its individual business direct customers.

By 2000, the company had formed its e-business group with the intent to make Intel a "100 percent e-corporation that maximizes profitability, responsiveness, and innovation." Success was to be measured in personalization, collaboration, standards, and return on investment. Along the way, the firm found it had to face some new challenges. It worked with the government of Taiwan, for example, to upgrade that country's network infrastructure to accommodate Internet activity. Taiwan is now one of the most advanced countries in the world in that aspect. The group had to deal with multiple languages so it turned to others for help. RosettaNet, the computer industry consortium for e-business automation standards, and other standards were developed to deal with these complications.

In May 2000, Intel offered the first release that focused entirely on enablement instead of automation to create a better customer experience. The WOM system had grown to eight applications and had to be rewritten to combine similar tasks and bring the number of apps down to five. Users were able to download simplified training modules for each application. An event notification feature was added to tell customers the status of their Hot List requests, change order requests, and return requests. Customers can receive notifications on the e-business Web site, via e-mail or both, in real time or batched according to their time frame. Later in 2000, the team implemented site balancing between customer locations, allowing customers to manage, balance, and assign inventory to multiple sites using the Internet tool.

The next step is to integrate the supply line management and Web order management systems. To do this, Intel is gathering data on how customers manage their inventory. Workflow analyses are being done as team members listen to customers make recommendations on how to help them manage inventories. After the optimal workflow is defined, software engineers will create a collaborative inventory management service that removes any customer tasks and steps that do not make sense.

Intel's experiences are a textbook example of how supply chain management can help manufacturers and their B2B customers increase revenues, reduce costs, and increase customer satisfaction. These partners can reap the rewards as they move from process automation to customer enablement using the new Internet technology as the tool of implementation.

Summary

We have gone through a mad ride as we watched the Internet appear, rise to great heights, and fall off precipitously, as would-be users misapplied its benefits. Now we are in a more sensible era in which the true benefits of Internet technology and collaboration will be felt.

It starts with understanding what you do not know about the Web and its applications and moving very carefully, with the help of partners, to find the applications that make the most sense for the firm's intended customers and consumers. Working collaboratively, these partners can share experiences, experiment with new models, and craft the e-commerce solutions that bring the potential of the Internet to the center of how business will be done for years to come.

9 Mistake 8: Lack of Collaboration Across the Extended Enterprise

Despite all of the rhetoric espousing inter-enterprise collaboration, and our argument in the preceding chapter, the major focus for most supply chain efforts today continues to be on making further improvements to internal excellence, regardless of the effect on external partners. The quest is to get as good as possible at all of the process steps within the four walls of the organization. Many firms make great progress with this intent as they build their internal best practices, but they overlook many of the better external practices or entirely new innovations that have meaning to customers. Motorola became very efficient at making analog cell phones, but was left at the gate when it came to the shift in demand to digital units.

The advanced approach, once a firm understands the true value of e-commerce, is to move externally, with the help of supply chain partners, toward being part of an active network sharing information on best practices, often of a global aspect. ERP and CRM tools have helped many firms build a foundation to make further improvements. Now these firms are seeking ways to get to the next level of improvement by sharing data, applications, and best practices with other supply chain constituents, often through their ERP systems. According to an *Information Week* research survey, "more than nine of ten business and IT executives believe collaboration — the sharing of business information within and across corporate organizations — will increase sales opportunities, and about half say it will cut costs" (McDougall, 2001, p. 44).

As firms begin to realize the hidden values residing in Internet technology and collaboration, they typically attempt some sort of pilot operation with one or two valued and trusted partners. The purpose is to determine the extent to which collaborative efforts, particularly those employing cyber technology, can enhance business performance. These are not easy alliances, either to form or to make them work. They require an enormous amount of dedication, planning, and support, but they are the steppingstones to the next level of progress.

As a sequel to the previous chapter, we will now take a look at the next supply chain mistake: lack of collaboration across an extended enterprise. It is a mistake that results in longer lead times, higher total costs, less customer satisfaction, and limited revenue growth. Before proceeding, however, a caveat is in order. Examples of successful networking, particularly on a global scale, are very limited. Such activity is really on the front edge of extended supply chain efforts. For our purposes, we will use those action stories we have to illustrate how a few pathfinders are moving their supply chain networks to the forefront of their industries.

Benefits Come from Sharing Information Across the Extended Enterprise

As firms try and extend their network and build in the next round of improvements, we find the major activity is in the business-to-business sector. Indeed, this area of attention continues to grow dramatically. By one estimate, during the year 2000, the total investment in B2B e-commerce infrastructure exceeded $200 billion (Agrawal, 2001, p. 22). We can expect that number to increase in 2001 and accelerate thereafter. Companies are not abandoning the Internet or the viable Net marketplaces and exchanges that could benefit an industry player. The confusion surrounds how to move forward to take advantage of Web technology and collaborate effectively across an extended enterprise. In spite of the fact that nucleus firms like Dell Computer and Wal-Mart can develop a competitive advantage from their exclusive collaborations and from the proprietary sharing of information with key suppliers, other firms refuse to get into the act. A little enlightenment might help.

The real benefit from using the Web will come as firms link parts of their organizations and seamlessly integrate information flows and work processes across their end-to-end network. The ability to speed up the flow of information and make it more widely available will cut development time, bring products more likely to succeed to market faster, and, in general, produce gains more than sufficient to cover the necessary investments in people and capital. The use of B2B exchanges, in concert with ERP and decision support systems, will be in the center of most of these situations. With supply chain costs now pegged at about 10 to 15% of revenue in most industries, the potential to take that amount down by two to five or more points is the next frontier for business management. But it now requires the help of network partners. That is what Dell and Wal-Mart learned.

A retailer, for example, which has significant demand for its products, still must depend on a supply chain that extends from the raw materials being delivered to factories, some as far away as Sri Lanka, China, Korea, and Taiwan. The goods most in demand must then be shipped to the distribution centers, from which they will be parceled to the stores. Getting to remote locations could involve an important distributor. Priority needs might be accommodated by airfreight. Getting across borders requires skill in customs and duty applications. It is simply becoming an integrated world, and the firms that do not build the extended enterprise infrastructure to support their supply chains are going to be left out of the game. E-commerce affords companies the opportunity to create or enhance the necessary linkages that make inter-enterprise efforts a success.

Consider a very simple example, on a much smaller scale than global, of how you can enhance a relationship and build value together. Home Depot announced such an effort when the home-improvement retailer began allowing employees from such suppliers as Georgia-Pacific Corp. to work the floors of the retailer's stores. G-P employees log into the store and let Home Depot track their time, sales, and inventory information, as these employees help customers in the lumberyard part of the store. The idea is to provide Home Depot with details on how some of the top suppliers perform when the supplier's people are brought into the stores to help. "Hopefully, we'll both end up managing a better business together," says Home Depot CIO Ron Griffin. Beyond its pilot with G-P, Home Depot is also sharing real-time point-of-sale information that aids suppliers to lower inventory costs; 85% of all the company's dealings with suppliers are now conducted electronically (McDougall, 2001, p. 43).

And we must not overlook one important point about this environment. As the concept of extended enterprise collaboration becomes more common, some firms will definitely be a part of multiple supply chain networks. As one business analyst has put it:

> In men's apparel, a retailer could have a number of supply chains. One might replenish perennials such as undershirts, white dress shirts, and size-40 regular navy blazers, while a second might stock fashion items

for which demand varies according to the season, the effectiveness of efforts to promote them, and their inherent appeal. A grocery retailer, meanwhile, must manage the flow of perishable produce (such as lettuce and apples) for which demand tends to be fairly predictable, and of non-perishable products (such as soft drinks) for which it can be influenced by heavy promotion. (Agrawal, 2001, p. 23)

Companies adept at meeting the needs of these business customers are going to be in demand by more than one nucleus firm. As they cooperate with firms and build network alliances, that does not exclude them from helping establish customized solutions for a number of nucleus firms.

Once a firm has determined its role in an extended enterprise network and decides to take advantage of collaboration, it sets about deciding how far to go and with whom. Let us look at an action study reported by Paul McDougall, for *Internet Week* magazine:

Last year Quaker Chemical Co., the $267 million chemical supplier to the steel and auto industries, decided to implement 'a flexible information-sharing network' that CIO Irving Tyler envisions as the basis for a loop connecting employees, customers, and suppliers. It is the kind of broad effort at collaborative business that draws from and blurs traditional technology disciplines from knowledge management to ERP and CRM.

The Conshohocken, Pennsylvania company's business-intelligence network, which operates over a frame relay WAN powered, in part, by software from Intraspect Software, Inc., lets employees and business partners throughout the world share product data, research, and other information in almost any format. The multinational company is using its network to more efficiently solve problems and take advantage of its global research network. 'Our associates in China can tackle a problem so that it's solved by the time our people in Europe get to work in the morning,' says Tyler.

With that backbone in place, Quaker is trying to develop collaborative networks internally and across its supply chain and customer base — and ultimately link them all together. If Quaker can connect the loop from suppliers to customers, Tyler sees ways to make the entire process more efficient. For instance, some of the chemicals that Quaker sells contain an anticorrosion agent because that is how the base ingredient comes from the supplier. That is fine for the steel mills, but the automakers have to remove the agent and apply their own formula. If Quaker knew its chemicals would be used on steel bound for the auto industry, it could use different ingredients. 'If we can push that information through the development chain, we could speed up our development cycles and create some new opportunities,' Tyler says. (McDougal, 2001, p. 48)

The trick is to integrate process flows across the extended enterprise. And it is happening in lots of places. Briggs & Stratton built a collaborative extranet, BriggsNetwork.com, for partners and suppliers. The eight-language site lets suppliers and manufacturing customers log in and check specifications, view sales promotions, and receive parts and warranty information. Diesel engine manufacturer Cummins Inc. has constructed an extranet that customers such as Peterbilt Motors Co. and Kenworth Truck Co. can access for updates on engine orders. Truck makers can also log into the site and view early prototypes for Cummins' 2003 line of engines.

Penske Logistics developed an extranet, through which customers such as cabinetmaker American Woodmark could access data to do their own monitoring of the exact comings and goings of the Penske delivery trucks. Trucker Schneider National Inc. created Schneider Logistics to market logistical data to its customers over the Internet as part of a larger logistics services offering. And Whirlpool Corp. in Benton Harbor has formed an internal unit called eWhirlpool to help push large retail customers away from proprietary EDI systems, which have limited functionality and are more costly to operate. Retailers using Whirlpool's system can use a Web browser to share order-management information, as well as gather product information. These are all harbingers of what is to come when firms sit down together and extend their reach across the extended enterprise and help each other find the next level of improvement.

Buyers and Sellers Must Evolve into Collaborators

To get to this next level of progress, it takes more than a change in attitude. There must be an alteration to roles played out by the participants. The traditional roles of buying and selling must move beyond the negotiating process and enter into real collaboration. That is done best in the beginning with a small number of key suppliers and buyers, working in an open atmosphere, and sharing proprietary data that helps each firm improve. What we are considering is how to get to the next level of supply chain enhancement.

If collaboration is all about working together to find solutions and the means to build revenues, then what is needed? Three types of activities are required:

1. Partners must connect their information systems to provide visibility to critical information that directly impacts the process steps in their connectivity. That means one or two suppliers come together with a

manufacturer, or a manufacturer meets with a key distributor or retailer, to construct a flow chart and begin analyzing where this visibility is required. This is a short-term or early-stage effort, but it is crucial to get started on the right track. Discussions will be tentative at first, but inevitably lead to getting valuable data online so people can access it across the involved firms. We will consider this type of effort when we describe a partnering diagnostic laboratory as one means of early collaboration.

We have found such efforts are best led by a nucleus firm having the scale, branding, and resources to act as the hosting firm. Such a company brings the partners together, organizes the preliminary meetings, and provides much of the impetus behind creating the enhanced processing. During this first stage, the parties work together on such activities as standardization so specifications and nomenclatures can be viewed without similar items being described in dissimilar manners. Procter & Gamble has done this effectively with a number of suppliers, using specific products in their detergents and oral care line as examples for standardizing and simplification. Tide had reached the point where there were so many optional packages, for example, that the consumer was being confused. The product could be bought in liquid or powder, from small to gargantuan sizes (try carrying the largest liquid container from Sam's Club), plain or with additives, until shelf space simply would not accommodate all the possibilities.

The team worked with specific customers and the suppliers to find the right number of reduced SKUs making sense for the consumer.

The parties discuss the means by which ERP systems can be linked in spite of the disparate systems used by different partners. The flow charts are reviewed in earnest to determine exactly what data can enhance the interactions. Planning, scheduling, and delivery information is shared so each party can become more efficient at its process steps. At all times, the focus is on what information will make the network more effective in the eyes of the customers and consumers.

2. Partners in an extended supply chain must map the sequence of steps occurring in the end-to-end linkage so the logistics and coordination process become seamless and as effective as possible. This is a midterm or intermediate stage effort and will have dramatic impact on the costs and cycle times involved in delivering the products and services expected by the targeted consumers. With advanced networks moving to virtual logistics systems, this has to be a part of any collaborative effort.

During this stage, the partners are looking at how they can share best practices and applications across their network. Attention moves to inventory management, so they can minimize the total investment, even if one party has to increase its stocks. Just-in-time efforts move to a higher level as the partners work on sequencing so the right goods are at the point of need at the right time. Tracking the movement of goods becomes a real-time event. Today, Toyota can track a part for any model car anywhere in its supply chain with an online system.

3. Partners must begin construction of a consumer-focused strategy that will introduce the kind of differentiating factors in the minds of the targeted groups. Inherent in this strategy will be the financial commitments, joint investments, cost analysis and risk sharing necessary to make networking a success. Attention to network-level decision making is important in this longer-term, advanced stage effort, to make certain risk is analyzed and shared among constituents, and a means of sharing in the results is clearly articulated.

Now the partners begin looking at budgets to cover the cost of the implementations being recommended by the first two groups. Decision rules are established to cover how joint investments will be made, how resources will be allocated, and how benefits will be shared among the constituents. If a special packaging machine will allow for the insertion of coupons or advertising material for a particular product, for example, and both manufacturer and retailer will achieve more sales, it could be a wise move to share the cost of the machine. If a supplier can run a machine better than the manufacturer, why not consider having the supplier's people run the operation? And always, how should the benefits be shared? Fifty–fifty is a common choice among constituents and is characteristic of the most advanced of supply chain networks.

Each of these steps should be carefully analyzed and executed while being coordinated by an overall steering committee. As mentioned in Chapter 2, there must also be a system of measurement developed so the partners can tell how they are making progress and when they have achieved success.

Exhibit 9.1 illustrates where most firms start when they embark on an extended enterprise effort. Anticipated performance does improve over time as supply chain initiatives are executed. Working to enhance internal excellence, most companies can show some impressive results, particularly through reengineering and ERP efforts. Virtually every firm we have studied, however, begins to show a slowing of the returns on the effort. It is

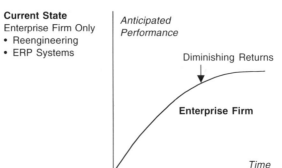

Exhibit 9.1 Current supply chain state.

not that the firm does not have initiatives to pursue; it just tends to begin showing lower benefit for the amount of resources applied.

In Exhibit 9.2, we see the concept behind extended enterprise collaboration is to build a second improvement curve on top of the slowing returns from current efforts. As companies come together and share the best from both organizations, the slope of the improvement curve has to increase. Now the enterprise and its selected partners benefit from a "lift" in results, as two or more firms begin to use best practices and collaborative commerce to enhance network optimization.

The change that is mandated is for buyers and sellers to both think more strategically and determine how they can go beyond price discussions to consider value adding features, which normally elude the typical

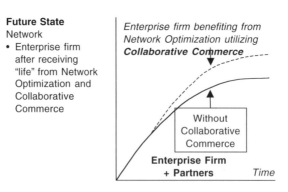

Exhibit 9.2 Future networked supply chain state.

purchasing activities. Now the focus goes to bringing more functions and people into consideration of how to build further improvements. Operations, logistics, data processing, marketing, and other departments are represented as the opportunity to have direct contact with customers and suppliers, previously not possible, enhancing the possibility to find hidden savings in the relationships.

Larry Lapide, VP of Research Operations and Business Applications at AMR Research, describes the transformation being considered from an e-commerce perspective. Most of the discussion and early implementation of inter-enterprise electronic trading partnerships have focused on business-to-business e-commerce through the automation of transactions using EDI. "This is not the same as collaboration; a higher form of e-commerce involving joint planning and scheduling," says Lapide. Traditional partnerships evolve in three stages, often starting with a transactional stage and moving toward information sharing. "Collaboration, the third stage, will usually follow, building upon transactional and information-sharing infrastructures" (Lapide, 2001).

Lapide believes that as the transactional relationships evolve, the need for new attitudes and roles becomes a factor in success. "While information sharing relationships go a long way toward enabling supply/demand synchronization," he explains, "they do little to help reduce the uncertainty faced by trading partners in determining future product supply/demand." To better enhance a buyer–seller relationship, some trading partners are moving toward more collaborative relationships. These efforts enable partners to work together to gain a better understanding of future product demand and put more realistic plans in place to most effectively satisfy it. "In the case of working collaboratively on consumer requirements," Lapide remarks, "trading partners might work jointly on new product designs and forecasting consumer demand. Inter-enterprise collaboration, while in its infancy, offers the most potential to drastically improve supply chain performance" (Lapide, 2001).

As mentioned, getting buyers and sellers to this new level of cooperation is not an easy task as the years of ingrained mistrust and desire to get the most advantage in every negotiation supersedes the idea of mutual resources being applied for mutual benefits. Experience has taught us that overcoming this inertia is best done with focused pilot efforts, one of which has been especially beneficial — the partnering diagnostic laboratory (PDL).

Partnering Diagnostic Laboratories Are a Tool for Starting Collaboration

One of the most effective tools we have applied to help firms get started on collaboration is a PDL. A PDL is a simple but very powerful means of bringing would-be partners together to analyze the relationship and the steps necessary between them to satisfy the ultimate customer or consumer. Having applied this technique with dozens of partners, we can say it has never failed to improve the relationship and bring attention to values not normally developed in the normal negotiation process. It has also always resulted in the introduction of actions that enhanced the supply chain network and returned far more than the investment in the effort. Exhibit 9.3 illustrates the technique being considered.

The PDL provides a focused workshop environment to identify and prioritize opportunities that bring increased value to supply chain constituents. It is designed to be a simple means of bringing partners together, either suppliers and manufacturers, manufacturers and distributors, manufacturers and customers, or any combination of firms in a value chain network that want to enhance their processing and take advantage of e-commerce capabilities. It comes in two phases and begins with a planning session, usually conducted with the nucleus firm that acts as a sponsor for the effort. The objective is to develop ideas and concepts to be pursued in the PDL and to build the framework for the effort.

As part of the planning phase, the technique is fully discussed, several executives are interviewed to get their perspectives on how the relationship might be enhanced, and a set of preworkshop hypotheses are developed. This one-day activity focuses on what the expected results should be and, particularly, which firm is an appropriate candidate for the PDL. Selection is best facilitated with outside help to make certain the firm meets some predetermined criteria for participation, especially to make sure the first effort is a success. With such a partner, attention will be focused on how to develop ideas based on best practices and how to consider specific industry practices, internally and externally. The intent is to achieve extra value in such areas as: cycle-time reduction, pooling of purchases, reducing dependencies on safety stocks, better inventory management, inter-enterprise communication systems based on the Internet, better asset utilization, aggregated transportation, and online visibility to data.

A decision is made to have a similar preliminary meeting with the selected partner or not. Most nucleus firms opt to have this additional session, so both parties come in prepared with a list of expectations from both sides and

Mistake 8: Lack of Collaboration Across the Extended Enterprise ■ 119

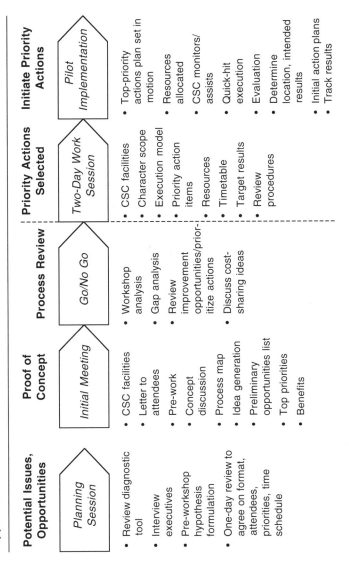

Exhibit 9.3 Partnering diagnostic laboratory.

what the potential benefits for both companies might be. With the partner selected, details are then worked out regarding how to create a simple, but meaningful, process map showing the product, data, and financial flows between the organizations, all the way to the end customer. The parties then determine the scope, purpose, details, and deliverables for the PDL. An effort to develop a preworkshop hypothesis (i.e., a statement of what improved set of conditions could be achieved) is usually included in this phase.

Toward the end of the first session, a gap analysis is developed to make a rough estimate of where the two firms might be in terms of significant measures, such as cycle times, inventory levels, transportation costs, transaction costs, impediments to performance, returns, and so forth. The idea is to get a feel for the potential benefits that could accrue. Then a decision to go forward or not is made and a list of invitees and letter of invitation are developed.

In the second phase, a 2-day diagnostic takes place. Now the objective is to develop mutually beneficial actions that will enhance supply chain processes and practices. The process map is fully discussed to make certain the important areas of interaction are included. The audience includes all pertinent functions from both partners. With consensus on the process flows, the group begins in earnest to consider all of the preliminary data from the planning session and begin constructing a strawman with ideas for an improved state of conditions. Brainstorming techniques are used to get active participation and generate ideas not normally considered by the parties.

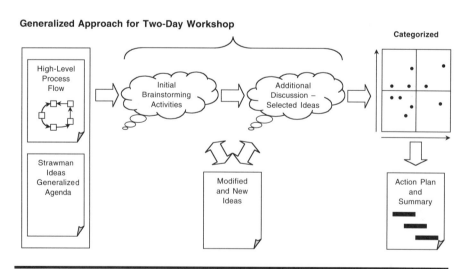

Exhibit 9.4 Two-day PDL workshop.

Through several iterations, the activities continue until a preliminary list of improvement ideas has been generated, discussed, grouped into categories, and thoroughly evaluated. Typical items on these lists include establishing a better communication system online, shortening the cycle time from idea to approval, reducing transportation and delivery costs, cutting the need for inventory, finding short-term cost savings, and so forth. Now teams are formed to analyze the top five or six actions that might be taken. These groups report on what they believe to be the cost and benefits for pursuing such efforts. When consensus is achieved on the very top actions that the group believes should be taken, the activity moves to developing specific implementation plans, complete with action sponsor, scope of effort, required resources, preliminary action steps, timetable for execution, and order of magnitude costs/benefits.

The actions are then arrayed on a matrix of some sort, usually broken out by time to implement and relative value for the effort. That means some sensible framework is used to set the actions into place by order of time and priority. A final report and summary of activities is then prepared for management review. The deliverables include an outline of potential improvements and opportunities to enhance the relationship, a plan to validate the concepts and develop actual test results, and action team formation. With management support and endorsement of specific teams and actions, the effort really picks up momentum and pilot efforts are initiated.

From previous efforts, the potential benefits include the means to:

- Reduce errors in order processing
- Lower cost of procurement, particularly in areas of nondirect materials and services
- Reduce order fulfillment costs
- Shorten cycle time from order-to-delivery and order-to-cash
- Improve cycle time from concept to commercialization of new products
- Lower inventory and safety stocks
- Lower warehousing and distribution center space and costs
- Reduce freight costs
- Introduce customer-centric metrics
- Develop e-commerce interconnectivity features
- Improve communication systems and the value of shared data
- Develop new, profitable revenues

In the following action study, we will look at how two firms came together to flesh out the activities described and build a better and more profitable relationship for both companies.

Action Study — The Construction Company

The Construction Company is a $5 billion firm that builds everything from apartments and sports arenas to high-rise office complexes and shopping malls. It is a reasonably profitable construction company that depends on subcontractors for completion of any project it undertakes. Relationships with these suppliers vary but are often contentious as the margins on each project are very small and profit depends on watching every dollar of expenditure.

The Chief Technology Officer decided an opportunity existed to enhance this firm's capabilities if it could develop an improved working relationship with some of its key suppliers. He determined a PDL could be the vehicle for attaining that purpose. Working with Computer Sciences Corporation (CSC) as the facilitator of the PDL, he began by organizing the preliminary session and inviting members of his staff, operating personnel, purchasing representatives, and information technology associates to join the activity.

During the first session, a flow chart of several typical construction projects were considered and a decision made to use a high-rise (15-story or more) building as the center of focus. A process map depicting the current conditions was then developed and used to consider where the processing could be improved. As the critical path to completing such a building was discussed, it became apparent that one of the most important partners was the firm selected to install the elevators. Following another lengthy discussion of what constituted a viable PDL partner, an elevator company was selected as the potential partner for the PDL.

The group proceeded to contact the elevator company and determined there was interest to participate, but only if the construction company would listen to recommendations and there was something of value in the effort for the elevator company. The CTO agreed to both stipulations and an agenda and letter of invitation was prepared for the PDL. Contained within that letter were the scope and expected deliverables of the 2-day session and a specific list of potential benefits for both parties. A time and place were selected for the PDL and each attendee was asked to come prepared to answer a few questions about the relationship and what each firm wanted to accomplish to help get the interaction started. Both firms agreed to come prepared with their latest version of a flow chart depicting the construction of a highrise building. With agreement from both firms and a list of the possible improvements, a go-forward decision was made.

At the two-day workshop, the group was polled to get the answers to their questionnaire. These answers and each participant's objectives for the sessions were listed and used to later determine if the effort had been successful. The next 8 hours were devoted to developing the preliminary improvement ideas, which reached an astounding 80+ suggestions. Following an informal dinner, where the parties really became energized and shared ideas that would never have surfaced in the normal processing, the group returned to the workshop and began prioritizing what they had discussed.

After considerable discussion that included consulting other people within both organizations to get information on the potential benefits of some of the proposed actions, the group agreed on six specific initiatives that they believed had the highest mutual benefits. Each action was developed so that a specific sponsor was identified, a preliminary scope and charter was created, and both firms committed resources to the action steps. Timetables were established along with how the subsequent pilots would be structured and results reported. At the end of the workshop, the participants agreed the PDL had exceeded their expectations. But that is not where the story ends.

Following the workshop, teams from both companies have met numerous times to pursue such activities as reducing the cycle time for constructing a high-rise building, improving the communication flows between the firms, bringing the elevator company into earlier involvement with drawings and specifications, and even helping with bidding the new jobs. Another action included improving the working relationships with the architects, designers, and building owners to help reduce time for construction and total project cost. One effort has been particularly beneficial.

Working together, the firms have developed a new "fast track" model for such projects. Using heavy doses of information transfer (i.e., early and often sharing of drawings, specifications, and engineering changes), expedited by the newly established and custom designed extranet linking the firms, the partners now approach building owners and designers with the concept of bringing their project to completion in 25% less time and with the possibility of considerable cost savings. Accomplishing fast track projects is possible by the dramatic improvements the group found to reducing cycle times and improving data transfer between the firms.

Summary

Once a firm discovers the Internet and realizes its true potential, it can begin working with supply chain constituents to use Web technology and collaboration to enhance the linked processing. Collaboration across the extended enterprise starts with selecting willing and trusted allies with whom the process steps in the end-to-end network can be analyzed and worked on for improvement. It proceeds toward finding enhancement of value to the intended customers and consumers.

While there are limited success stories for such efforts, we have considered both a framework within which they can be developed and a tool for initiating such alliances. The secrets to success lie in determining what the mutual benefits might be, how to share best practices across enterprises, and how to initiate e-commerce activities that lead to a new business model that differentiates the network in the eyes of the final customers and consumers.

10 Mistake 9: Weak Global Concepts

Buying, selling, making, and delivering goods around the world have become an everyday event for many firms as they take advantage of enlarging populations, modes of transportation, low-cost labor, and global demand to enhance a firm's revenues and profits. Much has changed, however, since a merchant named Marco Polo probably introduced the concept of global trading.

With proper preparation and use of the Internet, it is now possible to roll out products, not just in the most lucrative domestic regions, but also simultaneously across many foreign markets. As cyber-based trade is expected to hit somewhere between $6 and $8 trillion by 2004, depending upon which forecaster is used, businesses are hard at work trying to extend supply chains in an increasingly complex environment. It is now feasible to be a buyer, supplier, seller, or manufacturer in a B2B2C network anywhere around the globe. These aspects offer further opportunities to profit by extending reach to the supply chain.

Global Presence Comes with Headaches

Unfortunately, these opportunities come with demands on the supply chain — especially those of a language, culture, government, logistic, technology, and cross-national border delivery nature. An infrastructure that can handle everything, from receiving orders and dealing with local conditions to making fail-safe global deliveries, is a major part of any successful international supply chain. Achieving skill under such conditions requires a firm to meet daunting requirements.

Operating offshore facilities in foreign countries, for example, introduces a new dimension to a supply chain network, one that is often handled poorly. These factories, distribution centers, staging areas, or consolidating depots must be staffed with people as interested in the success of the supply chain as those who build and own them. They must function over many thousands of miles, as well as those who run facilities set up domestically, to deliver within 100 miles or less. Careful selection of partners and employees in these remote areas is a hallmark of those firms operating global networks successfully. It is also the scar that most failures bear.

Regardless of the complications, the move toward global facilities seems to be an unstoppable trend. A recent Forrester Research survey of 50 global manufacturing executives found that, "Global 2500 companies increasingly manufactured products all over the world." But this is not done without problems. The same report said that 75% of those interviewed noted inconsistent production systems and equipment. As one example, both U.S. and non-U.S.-based companies said, "relying on outdated forms of communication left them feeling disconnected from their supply chain partners. Thirty-eight percent cited poor visibility and 24 percent cited poor communications within the supply chain as their biggest problems" (Radjou, 2000, p. 30).

Many companies faced with these situations prefer to arrange alliances with other companies having facilities or the capability to establish viable operations in targeted foreign markets or areas of intended supply. Our experience indicates the leaders become masters at arranging such alliances. The less experienced seem to try and extend successful domestic models into foreign environments, only to recoil as they learn it is better to rely on trusted local partners. A few persevere at owning and controlling their own facilities and can point to very successful global operations that meet the needs of their supply chains. The necessity to prepare for such ventures and execute efficiently leads us to consider the next supply chain mistake — weak global concepts.

In this chapter we will delve into the growing desire to have a worldwide presence and the requisite need to back that desire with a viable supply chain infrastructure and delivery system that brings as much customer satisfaction as if the plant were next door to the consumer. Our look will be from three aspects: being able to source necessary raw materials or products on a global basis, manufacturing or distributing from offshore locations, or extending a firm's revenues by selling into foreign markets.

One caveat is in order before proceeding. The issue of global presence is a far-reaching subject and well beyond the scope of this text. We shall concentrate in this chapter on why there is such a lure for extending supply

Exhibit 10.1 The Demographics Cry Out for Global Efforts

Rank	Country	Population (millions)
	World's Largest Countries in 2001	
1	China	1,273
2	India	1,033
3	United States	285
4	Indonesia	206
5	Brazil	172
6	Pakistan	145
7	Russia	144
8	Bangladesh	134
9	Japan	127
10	Nigeria	127
11	Mexico	100
12	Germany	82
13	Vietnam	79
14	Philippines	77
15	Egypt	70

Source: Population Reference Bureau.

chains so far afield and what can go right and wrong as firms decide to expand globally — either to source, manufacture, or sell. The perspective taken will be that of a supply chain specialist looking for an adequate means of expanding revenues and cutting costs without incurring poor customer satisfaction.

Let us take a look at why people want to expand outside of domestic boundaries. It is not just a matter of expanding sales in new markets anymore. Today, the effort must be focused and part of a long-range plan that takes advantage of core competencies and shifting centers of commerce. In the first place, the U.S., Canada, Europe, and Japan may be very lucrative markets for the present, and are usually found at the center of most marketing efforts. These nations, however, do not dominate the population centers of the world and are far from the least-cost locales for labor.

As we consider population (or the number of potential buyers), there are factors at work altering the earth's demographics. Changes in birth rates, death rates, immigration, and other factors are redefining where people live. The latest census figures show that only 4.6% of the world population lives in the U.S., and only slightly more than 5% live in North America. The European Market is a strong center of consumption now, but is being affected by some of the alterations mentioned.

Exhibit 10.1 lists the world's 15 largest countries in 2001. The first two countries in that listing, China and India, make up 37.5% of the 6.1 billion people on our earth. That is too many people to ignore, especially if your firm makes and sells something that can be consumed in such locales or is seeking a lower-cost manufacturing site with a growing number of technically skilled people.

From the sourcing and manufacturing perspectives, the governments in these populous nations want their people to have jobs, and very attractive labor rates are often an inducement. To ignore such opportunities is to put the firm at risk if another foreign competitor moves sourcing of materials, subassemblies, parts, or finished products to such a locale. If labor content is a significant part of the cost of goods sold, most firms will have already looked at viable foreign sources. The leaders tend to work this possibility, as a key element in their supply chains, to be as cost competitive as possible. Any firm needing cutting and sewing of fabric, for example, moved to Sri Lanka, the Philippines, or Indonesia a long time ago. One caveat prevails, however. The supply chain requirement in these situations is to have an efficient system of delivery so the savings made from local production are not dissipated as the goods find their way to market.

From another aspect of size, many European countries are actually declining in population because there are more deaths occurring than births, a phenomenon not being experienced in other parts of the world. Russia is one country with the largest gap between births and deaths and is currently experiencing a decline in population of 950,000 people per year. As countries age, so does the likelihood that populations will begin to decline. Fifteen percent of Europe's population is age 65 or over, compared with 7% worldwide. Today, nearly all of the world population growth is occurring in the less developed countries. According to projections, only three of the more developed countries, the U.S., Russia, and Japan, will remain among the world's most populous countries by 2025.

Exhibit 10.2 ranks the 15 most populous countries, projected for the year 2025. Of interest, Indonesia, Pakistan, Brazil, and Nigeria will have over 200 million people. Ethiopia, Philippines, Zaire, and Vietnam will have more than 100 million. For all of the problems these nations have endured and will face, they will need to feed, clothe, shelter, and administer to their people. Those who acquire the money will want to spend it on luxury items. These are large and growing markets that will be supplied one way or another. Savvy companies are at work now building the supply chain infrastructure to maintain a viable presence in many of these countries. Health care is a particularly viable effort, especially for nations trying to stem the

Exhibit 10.2 World's Largest Countries in 2025

Rank	Country	Population (millions)
1	China	1,431
2	India	1,363
3	U.S.	346
4	Indonesia	272
5	Pakistan	252
6	Brazil	219
7	Nigeria	204
8	Bangladesh	181
9	Russia	137
10	Mexico	131
11	Japan	121
12	Ethiopia	118
13	Philippines	108
14	Congo (Zaire)	106
15	Vietnam	104

Source: Population Reference Bureau

tide of famine, malnutrition, disease, and general lack of what is considered basic care in the developed countries. Making the products in these countries can substantially reduce the cost of supply if solving the headaches is part of the effort.

Two Countries Show the Results Can Be Difficult but Very Attractive

One firm hard at work on such a global tact, and one that has learned from its share of mistakes, is Whirlpool. The Benton Harbor, Michigan appliance stalwart, which has opted for manufacturing in the countries served, has been developing a presence in the targeted markets of China and India for many years — as a manufacturer and seller. Initially, the firm made global forays, particularly in Europe, assuming it knew enough about foreign markets to jump in quickly and establish a presence — and do things the Whirlpool way. A number of local firms were acquired and switched to the Whirlpool methodology. In general, it just did not work.

As the firm approached China and India, they went through a transformation. In a country like India, where most people wear white clothes and need to wash them in water not up to American standards, the potential market is large for washing machines. But the average citizen would need to spend nearly a year's wages for an American-style unit. The normal domestic business model or the products it fostered did not fit in this country.

Through trial and error as much as strategy and planning, Whirlpool learned how to appeal to the Indian citizenry and make products they could afford. The firm made progress with its global strategy by investing first in understanding the societies, customs, political conditions, business conditions, and so forth, before establishing the infrastructure for building and selling its products. It used one learning experience in China to prevent a further mistake. In China, the company took a $294 million write-down in 1989, as it had to close two appliance plants. This occurred after the firm understood how quickly a developing nation would reach saturation on products only a few could afford.

Today, Whirlpool works with local distributors and members of its global manufacturing network to get the supply chain right. And it could not come at a better time. When domestic demand for large appliances is expected to be flat, Whirlpool is projecting its overseas demand will be up 17%. In India, the product line matches local needs and financial capability, and the firm now "uses local contractors conversant in India's 18 languages to collect payments in cash and deliver appliances by truck, bicycles, even oxcart." The results are impressive. Since 1996, "Whirlpool's sales in India have leapt 80 percent — and should hit $200 million this year. Whirlpool is now the leading brand in India's fast-growing market for automatic washing machines" (Engardio, 2001, p. 134).

From another aspect for potential sellers, the true buying power of the population has to be considered. In India again, the income per capita is only US $440, compared to $31,900 in the U.S. Not many people can afford U.S. products. But when that figure is adjusted for purchasing power (the result of lower costs for goods and services in the country), that number rises to US $2230. Making products that are scaled down for local conditions means a firm increases its opportunity for new revenues, in spite of what seem to be poor conditions. Procter & Gamble makes one kind of disposable diaper, the Omni, that is affordable and the leading seller in India.

When you stop to consider that the country is also steadily building a middle class that will some day equal the size of the entire U.S. population, you begin to realize it is a viable market to be considered. China is also slowly but inexorably moving toward building a middle class and the members of that sector will want foreign goods.

As we think of both of these very large countries, we conjure images of the pathetically poor for which we hope some form of relief will develop. At the same time, supply chains in those countries can be a sort of lifeline to bring in products and use the lower cost infrastructure to employ people — to export supplies, make parts, and sell products to the more developed parts of the world. If ever relief is to come to the masses in those sectors, jobs will be a key element, but they must be part of a supply chain network that moves effectively to the more prosperous nations. The underdeveloped nations simply need the help of other developed partners to build their own infrastructure and use the natural resources they have as they pull themselves out of poverty.

There are potential benefits on both sides of an arrangement with these populous nations. As China becomes a full World Trade Organization (WTO) member, for example, that country will have to accept the organization's protocols. Among the requirements are equal access to trade by reducing tariffs, elimination of import quotas and nontariff barriers, and easing restrictions for foreign investments. The WTO's rules also govern the settlement of trade disputes and enforcement of related penalties.

These factors will make an already attractive market even more appealing for sourcing, manufacturing, and selling. According to one expert in this area, "Once China enters the WTO, exporters and importers around the world will be able to fundamentally change the way they conduct business with China. Foreign manufacturers and distributors eventually will have full control over the storage and distribution of their products in China" (Gooley, 2001, p. 18).

This country looms as a major expansion opportunity for many firms but, consistent with the message being delivered, comes with its complications. "Logistics and inventory-carrying costs account for a staggering 24 to 34 percent of the total landed costs of toys, appliances, tools, and basic hard goods sourced from China," says one knowledgeable source (Gould, 2001, p. 49). Supply chain specialists will be needed to keep such costs balanced with the potential savings.

Distance Becomes the First Issue to Face

Even when a new market looms with great potential, or a foreign location offers great sourcing opportunities, the first problem is always the same. It is not in your backyard. The location from which supplies, materials, subassemblies, or finished goods are to be sourced might seem to be a small distance on a global map, but it can be a source of serious difficulty to a firm intent on providing above-industry satisfaction to customers. From one perspective, the more remote the source or the customer, i.e., the greater the distance, the harder it is to provide the usual personal attention to details and the more attractive local facilities become. From another perspective, the need to have elements of service quality anywhere in the world demands a firm create an infrastructure that can duplicate standards achieved in the best domestic markets. That can be a difficult task for some sectors of the world.

The issue of distance is also not measured only in miles or kilometers. The cultural differences between nations involved in trade will determine how the individuals interact. As one expert summarizes the situation, "Differences in religious beliefs, race, social norms, and language are all capable of creating distances between two countries. Indeed, they can have a huge impact on trade: All other things being equal, trade between countries that share a language, for example, will be three times greater than between countries without a common language" (Ghemawat, 2001, p. 141). Communicating effectively between global facilities becomes a serious challenge for any infrastructure.

Compensating for a lack of local presence is overcome in a variety of ways. One of the most productive solutions is to take the time to understand the local conditions and determine if there is a viable niche for intrusion by an outsider. With that information, the next move is to determine if a Greenfield operation makes sense or the more preferred solution, an alliance, should be made with locals already possessing the intimate knowledge of conditions in the targeted arena. A global freight forwarder, for example, can be extremely valuable in assisting with an understanding of the culture and the complications of bringing goods into and out of the country. They can give valuable advice in the design stage before you try to implement in unknown waters. They will also earn their fees as you determine if ships or airplanes are the right mode of delivery and they get your cargo and containers through the customs and duties along the way.

The logistical infrastructure must be a part of the resulting supply chain, and this factor varies widely by country. Ask anyone who has tried to get full truckload shipments made across Spain or move produce from Central American plantations to docks. In today's environment, firms must make certain the infrastructure in the selected regions is capable in multimodal forms — roads, seaports, airports, and railroads, with sufficient capacity for moving products across those modes. Dealing with special local conditions can also bring particular headaches.

Floyd Stone, director of global supply chain for Griffith Laboratories, a firm operating in 14 countries, relays one incident with their Columbian operation. "In the space of one month," Stone reported, "we had to cope with two port strikes, a 50-percent turnover in customs officials, and four changes in the documentation required by customs. We would not have been able to handle the situation in Columbia as well as we did without the expertise of our global freight forwarder and the in-country relationships developed by our customer. Networking with industry experts is essential for success in international operations," Stone advises (Stone, 2001, p. 10).

As a firm establishes either a manufacturing site, by itself or in alliance with others, or begins to source from foreign facilities, an effort must be conducted to map the supply chain system and make a thorough analysis of the costs at each step in the linkage. Time factors should also be plotted, as they are often as important as the costs when delays mean lost sales and poor satisfaction. Alternative means of transportation should be evaluated and a party familiar with all of the cross-border requirements should be present to advise on potential complications and help with solutions.

Above all else, as much consideration should be given to the cultural barriers as is devoted to the manufacturing and logistical complications. Historical and political associations shared by countries in the past can still have a positive or negative effect in the present. "Colony-colonizer links, for example, boost trade by 900 percent, which is perhaps not too surprising given Britain's continuing ties with its former colonies, France's with the franc zone of West Africa, and Spain's with Latin America" (Ghemawat, 2001, p. 142).

Different Standards Can Be Another Obstacle

With a determination that distance is not an inhibitor, the firm then looks very closely at the global site to be operated or used, and questions compatibility of systems and procedures with existing domestic conditions. Now the consideration gets more intense as the foreign and domestic ways begin to really conflict. As firms extend their reach for making, buying, or selling, they often encounter problems with local standards within the culture. When you take collaboration and technology on a global ride, you need to standardize the systems, languages, and equipment being used, or you end up with a hodge-podge of interference from disparate Web sites, languages, operating techniques, and high-priority initiatives. Delays in sorting out the confusion can kill the effectiveness of an otherwise good global network.

It may make sense to think globally and act locally in general, but the translation of that adage means the local people will want to develop their own standards and systems. They will want to do business their way. I learned this lesson the hard way when I was personally involved in setting up plants in Central and South America to manufacture boxes to pack bananas headed for Europe, the U.S., Japan, and other global destinations. Each country — Nicaragua, Panama, Surinam, and Ecuador — had well-intentioned and qualified people, all of whom wanted to operate, communicate, and measure results their own way. Without the help of well-connected agents who knew how to deal with families, governments, and contractors in the area, we would never have succeeded with the ventures. With their help, we established viable operations, worked out the differences in cultural norms, and were profitable within months of beginning operation.

Centralizing Web technology, for example, is generally resisted across multinational companies, in spite of it being a central part of modern business models. And the redundant investment in a multitude of processing centers can lead to millions of dollars poured into autonomous efforts. Unfortunately, the locals will want to operate and control their own portion of the network communication system. Hewlett-Packard discovered this reality as that firm embarked on a massive global expansion and now "estimates that 1,500 Web sites cropped up among its regional and business units over the past few years" (Wilson, 2001, p. 1). How do you deal with this condition? H-P has ordered some of its local units to rip out the products and tools they had purchased previously and replace them with technology that follows the corporate standard. It was the only way the firm could make the overall system effective and take advantage of its global reach.

The proposed solutions to these and other cultural problems comprise translations dealing in local languages, currency conversion, customs documentation, and tools that facilitate data exchange and supply chain processing. It mandates combining regional efforts into a common platform with a common data repository shared across the affected regions. And it requires knowledgeable people who can translate the information into something useful by those running the parent organizations. Once again, the best advice is to lay out a diagram of the expanding infrastructure and use qualified I/T specialists to help deal with the need for some measure of local autonomy while maintaining the efficiency of the overall system. With the proliferation of supply chain software that is occurring, for example, a firm with global intentions could spend years working to get the disparate systems collaborating together in an effective manner.

Distribution Costs and Cycle Times Differentiate Leaders and Followers

Knowing where to locate your distribution centers is another key component of a successful global supply chain. In international trade, the cost of manufacturing a product can amount to half or less of the total cost. Distribution costs, including handling, freight, warehousing, duty and customs, in-transit inventory, inspections, and returns, can exceed the cost to make the products. Distribution also takes up most of the total cycle time. Deciding where to locate distribution centers, consolidation points, and assembly and repair centers is vital to an efficient supply chain system.

Toby B. Gooley, senior editor at Logistics Management & Distribution Report, suggests that when supply chain managers put together global distribution networks, they should answer this question: Should we locate our distribution centers in the markets we are serving or should we set up a centralized DC outside of them? This question is pertinent whether the company is buying from external suppliers or serving external markets. She offers as an example of the importance of that question as one considers the necessities involved in conducting buying and/or selling in Latin America. In an area that includes nearly three dozen countries and stretches for more than 5,000 miles, she claims, "This burgeoning economic powerhouse has become critically important to U.S. exporters and importers" (Gooley, 2001, p. 17).

Drawing on her experiences in this sector of the world, Gooley has come up with some recommendations that supply chain managers should consider in their decisions. A top priority she recommends is "availability of reliable, reasonably priced transportation services." Regardless of the size of the orders and the cycle time for deliveries, you need to "locate your DC where the required transportation services are always accessible." Many firms shipping by air move their Latin American inventory to a staging area in the U.S. — Miami, Dallas, New York, and other locations offering multiple daily flights, including all-cargo services. In this part of the world, the larger cities have frequent flights, but the air traffic to the smaller cities can be very limited, especially where state-owned carriers are favored.

"U.S. exporters may be able to get low rates," Gooley says, "by providing backhauls for airlines carrying imports from South America (fish from Chile or flowers from Columbia)." From this aspect, it could be less expensive to ship products by air from the U.S. to Latin America, rather than ship air freight within the region.

The presence of other transportation services must also be considered. Ocean freight is an obvious alternative and many carriers and sailings are available to get products back and forth in the region, including between countries. "Hub cities like Montevideo, Uruguay," Gooley explains, "offer frequent service to adjacent countries. One U.S.-based company that has taken advantage of this service is Polaroid. This manufacturer of films and cameras flies products to a bonded facility in Montevideo's free trade zone (FTZ), then ships the goods from the FTZ by ocean each week to Vitoria in southern Brazil" (Gooley, 2001, p. 18).

Gooley sees another consideration having to do with whether a U.S. location would be preferable to a distribution center in the destination country in terms of meeting market demands while maintaining tight cost and inventory control. She reports that Lucent Technologies made the decision to collaborate with Danzas AEI Intercontinental, an experienced freight forwarder, to establish an export center in Miami. From there, "products and parts are shipped to in-country warehouses in the Caribbean and Latin America." Gregory Johnston, the Lucent executive in charge of logistics for the region, explains the firm's approach this way, "Lucent's logistics strategy is to position materials within our warehouse network based on customer requirements for installation, regardless of where the customer is located. Transportation is important, but what counts is a flexible distribution network using technology to streamline and improve the flow of inventory and management of the supply chain" (Gooley, 2001, p. 18).

For other companies not in the fast-paced electronic and high technology industry, shipping directly from a plant to a regional DC in Latin America may prove to be more economical. Large and dense loads tend to ship better in full container loads and are less expensive to move directly from factory to a warehouse in the designated country. But like all products moving from or into other countries, deciding on the supply chain infrastructure, especially facilities for holding and shipping products, requires an analysis of all the costs, cycle times, and service tradeoffs. All transportation options should be considered — availability of telecommunication services, power reliability, local import and export regulations and norms, and the stability of the economic and political climate. It is not the type of decision making that should be made casually, as there are too many ramifications, and today's business customer and end consumer is simply too unwilling to put up with anything they cannot find from a domestic source.

A Thorough Process Review Enhances Network Competencies

When a firm becomes convinced that a global infrastructure makes sense, in spite of the complications mentioned, it must construct its network as a competitive system that can match or exceed the capabilities of competitors. An analysis of what is currently available from other networks and an honest appraisal of strengths and weaknesses of the best competitor versus the planned network is a good first step. This evaluation must include the potential foreign locations and facilities as part of the greater network, from the sourcing, planning, manufacturing, staging, inventory management, distribution, and selling aspects.

With the influence of e-commerce being expected to dramatically impact multinational networks, two other considerations are how well structured is the network connectivity that will link the offshore sites with domestic locations and headquarters, and how capable is the network system that will be required to take advantage of any surge in e-business or electronic communications. Two sets of questions come to mind:

1. Is the supply chain linkage viable by existing (successful) corporate standards at each point of hand-off between partners? A positive answer means the efficiency of the global portion of the network will be as good as what is generally experienced domestically. If not, does the global linkage meet the needs of the targeted local market or provide a viable option for use in sourcing particular raw materials, parts, or actual manufacturing for the larger supply chain network?
2. Where any gaps do exist between acceptable standards and actual performance, what steps are required and in what time frame to achieve viable performance levels? That means a plan and timetable will be established that brings the facility to acceptable performance in a reasonable amount of time that does not exceed the committed funding and payback. Too many global efforts are abandoned because sponsors presented overly optimistic timetables for meeting the predetermined performance metrics.

As these questions are being considered, most global firms set up a methodology to monitor the progress and revise any planning before the effort becomes a bad experience. There are just too many stories of major firms experimenting with a global strategy that failed not to take the time for a careful planning exercise. Someone knowledgeable in total supply chain work, including the activity-based costs involved, should help establish that procedure and oversee results. Beginning with a solid process map describing the proposed end-to-end supply chain, this monitor should require a viable plan to be submitted for each crucial point of interaction. That means the sources of initial raw materials are identified and qualified, put to tests, and accepted as viable partners. The potential manufacturing sites are scrutinized and costs evaluated thoroughly to determine where the core competencies reside. Distribution is mapped, tested, verified under tough conditions, and determined capable of meeting competing standards.

Manufacturing competencies are put to an extremely rigorous analysis to make certain best practices will result in lowest total costs. This is a very difficult exercise for most organizations. As alternate foreign sources or manufacturing sites are considered, the local manufacturing group will typically insist they can match anything done offshore. I personally encountered one particularly difficult situation when working with a firm manufacturing home appliances. While the firm had a strong and generally effective cost improvement effort, there were areas we found where a foreign source was better, but not in the eyes of the domestic manufacturing personnel.

In one example, the firm was winding its own fractional horsepower motors. The superintendent of that area proudly showed me data confirming 12 consecutive quarters in which the plant had reduced it costs. Unfortunately, this area was a bottleneck in the manufacturing flow and an offshore source was available with 30% lower landed costs. In the end, senior management had to make the very difficult decision to close the domestic operation and use the foreign source, having proven its viability with a 6-month test and documentation of better results.

As the analysis and decision-making progress occurs, one aspect of global planning will begin to emerge — the need to only participate directly in the supply chain when there is a clear level of competency. Otherwise, a constituent must give way to a more qualified partner. As global supply chains mature, one thing has become apparent — the traditional linear supply chain operated by a single, centrally positioned firm will be replaced by a network system that could contain one or more constituents working with multiple networks because they bring skills not available from the central firm. These global value chains will revolve around the nucleus firms in the center of activities, but will be characterized by the presence of external companies, which can perform process steps better than any other entity. The smart nucleus firm will work this development to the point where fully owned and operated sites are challenged and kept in the system only if they have superior competencies. The final model will include supplying, manufacturing, and delivery specialists, all cooperating at Internet speed to satisfy business customers and end consumers.

In this scenario, some firms will determine that it is feasible to own very few assets. The era of the virtual organization conducting business on a global scale is already appearing. Some drug makers, for example, are moving toward outsourcing all of their manufacturing except for the production of active ingredients. Lucent has plans to outsource most of its global manufacturing, possibly in response to what one competitor, Jupiter Networks, has done. Jupiter does not own a single plant.

As this trend accelerates, we are going to see a wave of rationalization in the interest of supply chain efficiency. It will be a course of events impacted by the need for effective virtual networks and full e-commerce network connectivity across those networks. Navi Radjou, an analyst for Forrester Research, summarizes the situation this way, "Once offline specialists get their manufacturing assets networked, they will use their new-found supply chain visibility to identify and cut ties with slower-performing partners and use their network connectivity to establish dynamic trading relationships" (Radjou, 2000, p. 34). As these manufacturing connections develop, he sees further impact to existing infrastructures. With network connectivity, companies will "finally be able to evaluate their true performance in real time. But the picture won't be pretty for money-losing plants — companies will start trading their unprofitable manufacturing capacity on spot markets or liquidating those plants altogether" (Radjou, 2000, p. 36).

With the supply and manufacturing parts of the supply chain in order and meeting appropriate performance standards, the analysis turns its attention to delivery. The reviewers must pay particular attention to the shipment and logistics aspects of the total supply chain. Here the players are advised to consider building models that simulate the flows and inventories involved in global trading. Technology and software are available today that help this type of effort. Different scenarios can be considered with various sourcing, manufacturing, and distribution facilities; demands can be altered to determine how the simulated system reacts; and contingency plans can be added to strengthen the central model being used. The traditional silo approach to analysis has to be abandoned in this part of the effort. All of the players in the network must come together and rigorously determine if the proposed global system meets or exceeds both local and international requirements.

With a viable model working, the players then turn their attention to the supporting technology and communication systems. Linking ERP systems is usually a part of this exercise and a very daunting requirement. A global distribution system, however, must be integrated so capacity and delivery constraints are quickly made apparent and rectified. Flows of inventory must be visible online with high degrees of accuracy. Planning has to be visible from end to end so each player knows how to respond or revise its operations in the face of emergencies. The whole effort should not be rushed, but should be carefully analyzed, especially in light of the complications and solutions mentioned in this chapter. The results can be very rewarding, as we shall illustrate in the following case example.

Action Study — Hitachi Europe Ltd. and Exel plc

In 1995, new management in the Information Media Group (IMG) of Hitachi Europe reviewed their business and how it was performing. IMG manages the supply of DVD and hard disk drives, monitors, plasma screens, and LCD projectors to most of the major personal computer original equipment manufacturers (OEM) across Europe, as well as the repair and street markets. One of the major conclusions was that the existing supply chain would not sustain the expected business growth and needed to undergo a major redesign process. This transformation was seen as one requiring a special supply chain partnership. Hitachi could not accomplish its objectives alone. What was envisaged was a multiphase plan that proceeded through these phases:

- Phase 1: Get the Basics Right

    The objective became to get the existing supply chain linkage from factories in Japan and Asia, via a distribution center in Europe, to OEMs across Europe and make it work at above-industry standards. Many of the issues with the existing systems related to the need to cope with the day-to-day problems created by the inefficiencies of the overall process. These conditions prevented Hitachi operational management from dealing with the causes of those issues and the long-term requirements of developing a world-class supply chain network.

    It was also envisaged that, while the service provider engaged to help with the solution could be the long-term partner, this was by no means a foregone conclusion. Serious thought was given to that part of the decision process. Exel plc was eventually selected as the partner. The merger of Ocean Group plc, the parent company of MSAS Global Logistics, and Exel Logistics created Exel plc. The new unit offers global supply chain management solutions. It was a good decision, achieving predictability and reliability of service, allowing Hitachi management to focus their attention on the second phase of the effort.

- Phase 2: Improve/Automate Information Flows

  Hitachi quickly began to task Exel to deliver more than simple freight forwarding solutions. An early move was to develop innovative IT solutions for Hitachi's business customers. The priority was to link Hitachi's existing global logistics enterprise system to an Exel warehouse management system and begin the process of information integration. These systems were made to work effectively to support the initial growth of IED's business, but with the approach of the year 2000 and the need to move to an ERP-based platform, a new solution was required.

- Phase 3: Increase Visibility, Level of Detail, and Quality of Management Information

  With the introduction of SAP R/3, Hitachi moved into a different era of information availability and the previous developments were discarded. It became apparent during the planning phase of this part of the project that the traditional customer/supplier, or account management, relationship would not work for a project where success was so reliant on both sides delivering on their promises. For that reason, Hitachi and Exel set up a Joint Strategy Board with working groups covering key subject areas. Exhibit 10.3 illustrates the structure used. Each group was chaired by a member of staff from the company with the greatest interest or expertise in the area.

  For the Information Technology portion of the proposed structure, Exel, through its Insight (purchase, sales, and order management software) and Unison (warehouse management) systems, was able to provide line item and individual product detail for all products from export origin via the European DC and on to the end customer, with multiple search options. Radio frequency bar code readers in the warehouse allowed almost instant data updates into the system. These readers also allowed for improved product tracking as each product has a bar code serial number required for warranty purposes. In-cab satellite tracking also increased the security of Hitachi products in the Exel distribution operation.

  Frequent transfers of information between the Hitachi and Exel systems ensured, for the first time, the real-time availability of business-critical data within Hitachi. This capability allowed Hitachi time to begin planning its inventories from the traditional push operations out of the Hitachi factories into a more responsive pull system in line with OEM requirements and business models. As product life cycles

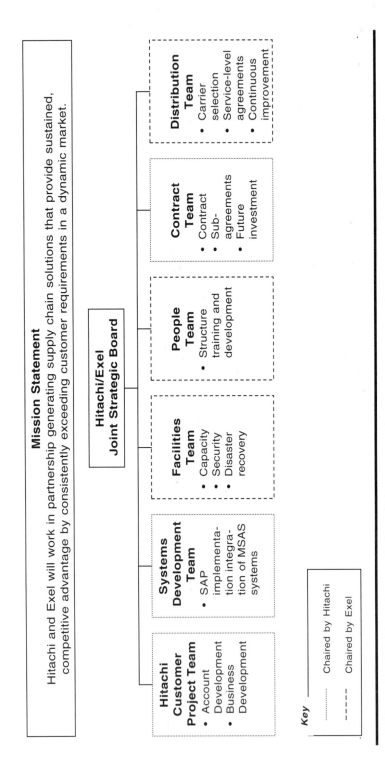

Exhibit 10.3 Hitachi/Exel strategic board.

were rapidly reducing from 1 to 2 years to below 6 months, this change was critical.

Unfortunately, the partners discovered that aspects of the technology introduced in Phase 3 were PC based, which limited access to a few people with the relevant software on their desks. As a result, although the third phase had given Hitachi access to a much higher level of detailed information, it remained in the hands of a small number of people who continued to handle all inquiries from sales divisions, specific customers, and so forth.

- Phase 4: Provide Wider Access within Hitachi — To Improve Information Availability and Increase the Flexibility of Operations

    Phase 4 saw the partnership move to focus most of its attention on the information requirements of Hitachi's business. As can be seen in Exhibit 10.4, the physical movement of product is essentially straightforward. Hitachi manufactures products in factories across Asia and Exel is responsible for collecting product from the factories, providing international transportation to warehouses in the United Kingdom (U.K.) and the Netherlands, and handles order fulfillment and final delivery to the customer. The information requirements for such an operation are now increasingly complex as described in Exhibit 10.5.

In response to the complexity of customer requirements such as those experienced by Hitachi, Exel has developed a Web-based response system termed supply chain integrator (SCI), which pulls together the information from its own enterprise systems *and* the systems of third party providers it manages for its customers. This feature ensures the system has the information management capability to handle these requirements. The implementation of SCI has achieved even more. It allows access to Hitachi data to anyone authorized within the company, anywhere in the world, and to anyone outside the organization, provided his access has received appropriate approval.

The result has been the removal of a huge amount of telephone and mail inquiries from within the organization, once again allowing the redeployment of internal resources to the task of improving services to customers and reducing the cost base for providing that service. Equally important, it has provided flexibility in the way that data are managed to allow Hitachi to change how it operates with key customers. In its constant effort to reduce inventory and improve customer service, Hitachi considered, with the help of its partner, the introduction of direct shipments from factory to specific

144 ■ The Supply Chain Manager's Problem-Solver

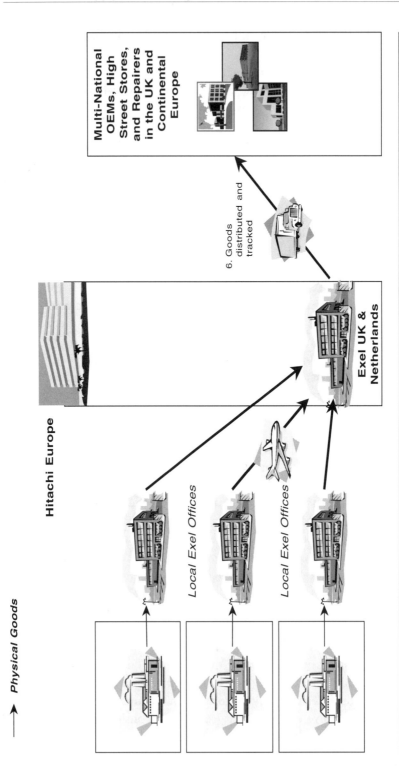

Exhibit 10.4 Physical flow of product.

Mistake 9: Weak Global Concepts ■ 145

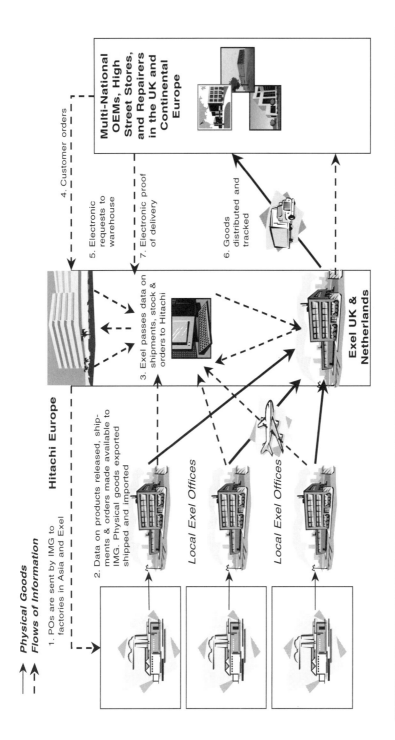

Exhibit 10.5 Information flows.

customers, cutting out the stop at the European warehouse and outbound logistic time and cost for such warehouse stops.

However, the traditional "drop shipment" was not acceptable. Hitachi wanted to introduce the direct shipment operation without relinquishing any of the visibility that currently existed. SCI is able to accommodate this requirement within its standard suite of applications. Exel now manages shipments from Hitachi factories in Asia to customers in Europe with full visibility throughout the transport movement, but without the products ever becoming a part of a European inventory. This operation went live in January 2001.

- Phase 5: To Increase Information Availability for Hitachi's Customers
  Phase 5 resulted in the production, by Hitachi, of a customer Web site that combines information produced by Exel systems and that is generated by SAP. As its development continues, it will become a single inquiry facility, allowing key customers access to data, enabling them to plan their own production scheduled based on component availability from Hitachi. Once again, the number of inquiries lodged with Hitachi by other mechanisms has reduced dramatically and the ability to interact directly with certain aspects of Hitachi's SAP system has given its customers unprecedented control and reliability in the crucial area of component supply.

- Phase 6: Farther and Deeper
  Phase 6 will allow Hitachi customers to place their product orders via the Web site and have the ability to download relevant information from that site into their own ERP system. This move will only enhance their ability to do effective planning. This information will move farther back up the supply chain, possibly as far as the Hitachi factory production forecast and demand management systems.

Summary

Global sourcing, manufacturing, and selling are attractive business options but are not for the novice. The field is littered with too many bodies of well-intentioned but unsuccessful businesses to overlook the inherent obstacles, which include culture, distance, language, governmental restrictions, and a myriad of logistics and distribution complications. A solid global network has viable players and efficiency at each point in the processing. It has a supply chain that operates as smoothly under international conditions as it does under domestic conditions. When we consider effective global networks, we typically see competent partners working in an international network, with focus on each process step, as an opportunity to differentiate the network in the eyes of the intended customer.

11 Mistake 10: Absence of Advanced Sourcing Applications

There is no area of supply chain management that gets more attention than the collective functions of purchasing, sourcing, and procurement. As a supply chain initiative is started, one of the first efforts is in this area that, for our purposes, we will term sourcing. The sourcing base is reviewed and generally reduced to a much smaller number of suppliers. Volumes are leveraged for best pricing and special features with emphasis typically on the areas of direct buying that most impact operations. With the usual limited number of resources, less attention is given to other areas of indirect spending and less critical materials and supplies. Attention in those areas comes later.

Our research shows virtually every Level 1 and 2 company involved in supply chain management has conducted a sourcing exercise and can document savings that have flowed to the profits of the firm. The amounts vary but are typically in the range of 5 to 8% of the total buy. As prices are lowered and suppliers shave margins on their goods and services, the effort does not end, however, as there is a relentless drive from senior management to continue to reduce costs through sourcing. With purchases representing anywhere from 25 to 75% of a firm's cost of sales, it is a natural tendency to want to keep attention focused on the supply base and those responsible for awarding the business. The result of sourcing becomes a central focus throughout the supply chain evolutionary process.

Two Concepts Drive Further Effort

An unusual set of conditions occurs, however, as firms continue the pursuit of cost reduction through better sourcing. Companies tend to split into a dichotomy. On the one hand, most firms insist that the sourcing group should be relentless in working the supplier group for further savings. Regardless of the improvements generated, the push is to keep moving the supplier's prices down. This type of effort is particularly pursued by senior management groups, which have little to no direct contact with the suppliers and lack an appreciation for the further benefits advanced sourcing techniques can bring to overall supply chain efficiency. A classic example of this approach was that fostered by Ignacio Lopez at General Motors when he insisted that his people find savings, even if it meant ignoring contracts and long-term agreements with suppliers.

From a second perspective, some firms seek further savings by working with at least a portion of the supply base in an atmosphere where other improvements are sought and savings are shared. This group approaches supply chain from the perspective that suppliers can help in introducing new processes and systems beneficial to both sides — buyers and sellers. Thomas Stallkamp exemplified this approach when he was executive vice president of procurement and supply for Chrysler Corp. While at that firm, he developed the idea of offering incentives to suppliers. Under his plan, if a supplier found a way to save money for Chrysler, the supplier kept half of the savings. When Daimler took control of his company and Stallkamp was dismissed, the firm went quickly back to the former concept and insisted that suppliers immediately slash pricing, giving up at least 5%. One report indicates that, under these circumstances at Daimler-Chrysler, "pricing is overwhelming all other considerations — in some cases even quality" (*Business Week*, June 4, 2001, p. 30B).

As a supply chain effort matures, regardless of which approach is fostered, those responsible for trying to squeeze as much as possible from the supply base run up against a pricing wall, which we call diminishing returns. At some point, even the most loyal and dedicated suppliers will plead for relief as there is a finite limit to how much price reduction can be given. After that point, the savings to the buyer come straight from profits for the seller, and most sellers will have no recourse but to refuse further price reductions. Those companies that remain anchored in Level 2 of the supply chain progression tend to fight this situation and will hammer away at their suppliers for more savings. In time, that becomes a fruitless exercise and most sourcing managers will turn to more beneficial techniques.

For firms that progress to Level 3 and beyond, the second approach takes on more meaning. Now the attention turns to how other aspects of the buyer–seller relationship can be leveraged to find nonprice improvements that still positively impact profits. This is not a difficult transition for some buyers who have worked for years with suppliers to find value-adding ideas and procedures. Using techniques born of previous value-enhancement efforts, new materials are analyzed, improved assemblies are tested, and novel transportation methods are evaluated. Inventory becomes an issue as the buyers and sellers look at how much is really needed, who should own it, and when it should be delivered. Order processing is discussed to see if transaction costs can be reduced. Most firms move in Level 3 from a mail, telephone, and fax system to an electronic data information (EDI) system. The leaders progress beyond that method in Level 4, to an e-commerce system that is fail-safe in operation and brings order processing costs to a bare minimum.

Working with a small group of key suppliers, most Level 3 and above sourcing professionals turn their attention to matters of a strategic nature. That means they focus on how the relationship can move to special features offering further advantage to both organizations. The parties look at the ownership of joint facilities like warehouses or distribution centers. They consider investments in joint assets, like packaging equipment or software, which shortens cycle time or speeds delivery processing. The idea is now focused on how benefits can be derived for both buyer and seller. Several of the more advanced organizations work together to build new revenues that benefit both parties by making joint sales presentations to targeted customers. That is a tough transition for many long-term professional buyers, especially those under executive pressure for pure price reductions.

In this chapter we will take a look at the more advanced sourcing techniques and see what some of the leaders have accomplished and why others have overlooked this opportunity. We will address the next supply chain mistake — absence of advanced sourcing applications. The thesis is very simple. Efforts to reduce sourcing costs and thereby positively impact profits are never going away, but improved techniques that take advantage of advanced supply chain procedures will provide greater results than a continual hammering at price concessions. The capabilities, for example, to automate negotiations, streamline actual procurement procedures, and manage planning and inventory movement through the Internet are just a few of the potential new tools that can bring the next dimension to collaborative sourcing.

R. David Nelson, vice president of worldwide supply management at Deere & Co., the Moline, Illinois equipment manufacturer, estimates the potential to reduce the total costs of goods and services in a supply chain to be in the range of 20 to 30% (Smock, 2001, p. 113). That is simply too much potential to overlook. But it is going to take some advanced techniques, and that will be the subject we consider next.

As buying organizations move their efforts under the end-to-end supply chain umbrella, they must be mindful of the earlier caveat we discussed — do not throw the baby out with the bath water. Most sourcing groups have been adept at helping any supply chain improvement process. As supply chain becomes the overarching focus of attention, care must be exercised not to abandon good practices, strong supplier relationships, and previous design and development work that were fostered with what is usually a core group of trusted suppliers. The work with this central supply group, however, must move from tactical issues of price, quantity, quality, delivery times, and terms to working together to create a strategic approach to sourcing that benefits buyers and sellers. Exhibit 11.1 illustrates, in greatly summarized fashion, the progression that is made as firms move from Levels 1 and 2, internally focused efforts, to the advanced techniques being advocated.

One of the first obstacles to be encountered on this trip will be another pressure typically foisted on the sourcing group — the demand by senior management to reduce headcount or the number of people doing the buying.

The Sourcing Progression

Levels 1 and 2 — internal focus	Level 3 — external focus	Level 4 — e-commerce	Level 5 — e-business
Volumes of buy are aggregated at a business unit level and price concessions are sought; some attention is given to transportation costs	Corporate-level buying occurs; best practices are applied; category buying moves to most capable buyer	Supplier expertise is sought and Web-based buying occurs; auctions, portals, and e-procurement techniques are tested and used	Best member of the supply chain network controls category; focused market places are used; consortium buying occurs; focus is on total network costs

Exhibit 11.1 The focus must move from tactical to strategic.

There is an idea pervading business today that says the number of people in this internal function can be cut dramatically because of the automation of the procedures taking place. Exhibit 11.1 does indicate that greater amounts of automation pervade the advanced levels in this function, but as will be explained, that occurs later in the progression and tends to work best with nondirect materials and services.

Sourcing executives are well advised to establish collaborative relationships with a core group of e-commerce-enabled suppliers before launching a buying Web site or participating in electronic marketplaces. That takes careful planning and development with the best of the sourcing professionals who have time freed up to work on this more strategic effort because of automation of routine tasks. The push for major reductions in force in sourcing is a dangerous concept as, more appropriately, the emphasis should be on upgrading the skills of this vital supply chain function, especially in the early stages of the progression.

When David Nelson joined John Deere in 1997, he added 175 people to his staff. More importantly, he made sure these people were placed in strategic supply positions and not forced to be tactical operators. His theory was that for every dollar invested in purchasing talent, you could gain ten dollars in profits. One hundred of his new associates were focused on supplier development and 50 on cost management. The remaining 25 were designated as best-practice specialists. Pay levels were increased and on-site training begun in earnest, with help from Arizona State University. Three years later, significant results were generated and other firms have adopted many of Nelson's initiatives.

From our perspective, none of the initiatives has been more important than moving the attention from getting jobs done, the tactical view, to finding ways to enhance the processing, the strategic view. As Doug Smock, editor of *Purchasing* magazine, sizes the situation, "If you, or the purchasing people working for you, are simply processing paper or orders, you are working strictly in a tactical mode. And your future is in question" (Smock, 2001, p. 113). Buyers must move from treating the core group of essential suppliers as adversaries to working with this group, often in an electronic format to find improvements benefiting both parties' supply chain. That begins the journey toward strategic sourcing.

In the beginning, as shown in Exhibit 11.1, the emphasis is typically placed on leveraging volume for price concession. That is an age-old proposition and one expected in the buying and selling arena. During this phase, however, as the number of suppliers shrinks, the number of larger volume key sources increases. When the dust settles on the new dispersion of the aggregated volume, a group of essential suppliers will appear. Now the sourcing function has a group with which to work on advanced techniques. Developing those techniques requires a major change in mindset, from viewing these suppliers as necessary adversaries to trusted advisors.

Suppliers as Partners Becomes the New Concept

For many buyers, this transition will not be difficult, as they have dealt on the basis of mutual benefit with a portion of their supply base for some time. It is the overall company attitude that must change. As we have stated, most of the pressure for unrelenting emphasis on cutting suppliers' prices comes from above the buying function. Gene Richter led a successful improvement to the purchasing effort at IBM. A key element in the transition was moving the mindset from suppliers as adversaries to suppliers as partners. According to Richter, "Everything at IBM used to be a secret. In procurement, we were the guardians of confidential information. You couldn't have effective collaboration with suppliers because IBM didn't want suppliers to know what product their part would be used in. What happened is we woke up. We realized that we couldn't be expert at everything" (Smock, 2001, p. 114).

In Level 3 of the progression, this wake-up attitude takes hold and a few of the most important, trusted, and cooperative suppliers are brought in for a discussion that quickly becomes strategic. Corporate-level aggregation of volume is used in this level, so a potential supplier/partner can look at the full volume being consumed across a firm, perhaps even including a few of the firm's other key suppliers. An office product's company, for example, could be given access to the full corporate buy and that of several key suppliers interested in the better arrangements to come from a fully aggregated situation.

An obstacle must be removed to move forward. This part of the effort is generally resisted much more by senior management than the sourcing group. Business unit executives generally resist relinquishing any control over any function, even if doing so means more profits for the company. The edict typically has to come from the CEO level to combine volumes in areas that have no differentiating effect in the market to get aggregated buying across a firm moving.

The Boeing Company faced this issue and moved aggressively forward, following its acquisition of Rockwell International and McDonnell Douglas. When the new organization confronted its purchasing requirements, it discovered the combined groups spent about $3.5 billion annually on nonproduction goods and services, as one example of how large the new sourcing would be. Categories included everything from office and shop supplies to machine tools, professional services, vehicles, computers, and software. Buyers were dealing with 17 purchasing systems across the 3 companies, all based on technologies from the 1960s and 1970s. Aggregating the information on total buy and sources was difficult and often inaccurate. Employees were also known to circumvent whatever contract or system was in place and paid full retail price for items.

In 1999, the company formed Boeing's Shared Services Group, which was organized to handle all nonproduction purchases. The group launched an enterprise-wide, Internet-based system for ordering, acquiring, and paying for such items. It was not an easy effort to implement and required senior-level insistence to get it going. According to Candace Ismael, director of supplier management and procurement, "The vision was to take all of those multiple, back-end purchasing systems and create a single process and supporting system for purchasing indirect parts" (Vijayan, 2001, p. 18).

As part of the new system built around Oracle Corporation's Internet Procurement software, thousands of Boeing employees worldwide can purchase and pay for office products from a Web page. The user connects to the appropriate Web page, searches an online catalog of preapproved items for which prices have been set, adds items to a shopping cart, and submits an order. User profiles determine the employee's buying authority and orders are routed to the appropriate manager for approval. Suppliers electronically bill Boeing and are paid via electronic fund transfers made directly to the supplier's bank.

With this level of support and direction, the buying and selling parties now begin to view the relationship, as Doug Smock, editor of *Purchasing* magazine, suggests, as one of recognizing the profit-expanding potential of purchasing and supply management. The key suppliers are gently prodded to determine where better buying practices occur. The question becomes: With which company do you have the best win-win situation? Techniques that can benefit both firms begin to emerge. Special categories of purchasing that do not directly impact the business unit or interfere with their market differentiation are identified. Such areas as travel, telecommunications, computers, software, office supplies, MRO supplies, etc. are brought under the control of a category specialist empowered to buy for the entire firm. This move has little to no effect on how different business units market and sell their products so agreement can be generated. With a key supplier, this category manager seeks out the optimum conditions under which the products can be supplied to the full aggregation.

As the firm and its core group of key suppliers increase their interest in the effort, they move to the collaborative phase of the progression — Level 4. Now the buyer seeks out the supplier's expertise in earnest, especially in areas not traditionally included in the negotiation process. In advanced sourcing efforts, the interface between buying and selling firms is greatly expanded. Information specialists are brought into the discussion, as well as logistics, distribution, operations, engineering, and other experts. Process maps describing the end-to-end supply chain are discussed, dissected, redesigned, and improved so both parties can reap benefits from the changes. E-Commerce features are very much a part of this phase of the collaboration.

One example of this type of advanced work is presented by General Mills, a firm that has been hard at work finding the next level of process savings in its supply chain. In one situation, General Mills has been working with a core group of packaging suppliers using cyber technology for packaging specification management. The ideas are to save money in materials by matching specifications with actual use and need, to share information technology practices, cut transaction costs, get successful products to market faster, and reduce costly production errors and the need for returns.

Kevin Fitzpatrick is leading this effort at General Mills, as he has moved from being the director of packaging to director of purchasing and strategic e-business alliances. At the center of this collaborative effort is a "Web-based tool called SMA/RT (for Specification Management Application/Real Time), established by Internet packaging company Empriva Technologies." General Mills plans to put all of its packaging data on this system, "allowing graphics professionals, designers, purchasing staff, quality managers, and other team members to communicate electronically through a common site" (Smock, 2001, p. 114). With this system, all the key constituents in the supply chain process steps can view a packaging project in real time, make improvement suggestions, take actions and view results, and give approvals online.

This example illustrates an area where progress can usually be made quickly through collaboration, but generally gets overlooked when too much emphasis is placed on pricing. Product development in the advanced mode of network operations cannot be done without direct collaboration of key suppliers. A team is usually formed early in Level 4 to look at how the cycle time from new concept to product commercialization can be reduced. Improvements of 50% or more are not unusual when firms link their design groups by sharing information across their computer-aided design and computer-aided manufacturing systems (CAD-CAM). Working with one consumer products company, which had historically lived with a 24-month cycle from idea to in-store sales, the time frame was initially cut to 12 months and, after a second effort, to 6 months. This was a win-win situation as suppliers were given a preferred position in the company's sourcing strategy and the buyer was able to generate significant new revenues in a shorter time frame.

In the most advanced area, Level 5, full network connectivity occurs. Now the most capable member in a supply chain network purchases goods and services for the full network members. That is a truly difficult concept for most organizations to embrace and, admittedly, it comes slowly to the involved firms. As supply chains mature, however, so does the amount of process steps that moves to the most competent constituent. Outsourcing grows as a reflection of attention to core competencies and total cost. Those companies approaching Level 5 find the percentage the firm makes itself will decrease and the amount sourced externally will increase.

Such efforts are not part of the traditional buying process. They involve alliances, often of a global nature, and long-term, mutually beneficial pacts that could involve co-investment in assets. These new arrangements require suppliers as specialists in particular segments of the supply chain processing. How far this type of transformation will go is not yet clear, as there is still a lot of testing, piloting, and new alliances underway. A few reliable analysts have predicted that some manufacturers could eventually find 80% or more of their costs will come through external sourcing as this trend increases. And it will not be done manually or with lots of face-to-face negotiation. The processing will be in some electronic format.

E-Procurement Is Altering Traditional Perspectives

Exhibit 11.2 illustrates where we see sourcing heading as supply chains enter the most advanced levels of progress. A nucleus firm will be at the center of the activity, as discussed in previous chapters. Around this nucleus firm will be activities that have typically fallen outside the sourcing function, but which we are now suggesting are very much a part of working with a key group of trusted suppliers, with the help and direction of advanced sourcing strategists.

On the left of the diagram, we see the usual area of sourcing expertise. The buyers are working with key suppliers to leverage purchasing capability, seek product discounts, and elicit custom offerings in exchange for commitments for long-term volume purchases. At the same time, the sourcing strategist is looking at the top of the diagram to find sources willing to work collaboratively on joint ventures, which could include electronic-based learning and safety programs that can be delivered across the network's extranet.

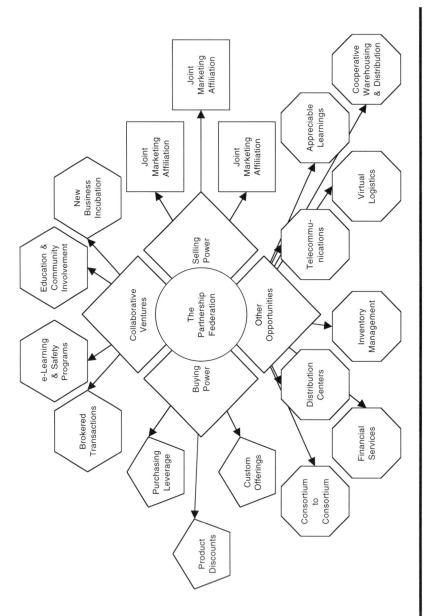

Exhibit 11.2 The Partnership Federation model. (Source: S. Horvath, Stesco, Inc.)

Some suppliers could be proposing joint ventures or new business incubations. Third-party organizations could be introduced to the conversations to broker an especially good warehousing arrangement or logistics plan that reduces costs for both firms. Software specialists, adept at linking CAD-CAM systems or reducing product development cycles, are often beneficial participants in this segment of the collaboration.

To the right of the diagram, we see selling power enhanced through such activities as:

- Joint marketing affiliations, where the firms work together on targeted promotions or new product developments aimed at a specific consumer group or to co-develop marketing campaigns aimed at those consumers. Companies participating in consortium buying could also be offering their products to other members of the consortium.
- Reciprocal marketing, where the cost of a marketing campaign and advertising is shared and suppliers have a say in how the effort is conducted.
- Customer/employee affinity programs, where joint use of telecommunication systems is used to open new sales channels for network partners or to give access to employees into products offered by the buying firm and its suppliers.

At the bottom of the diagram are other opportunities that include a potpourri of options that could benefit both organizations. Buyers and sellers could look at using distribution centers together. They could use the same bank for financial services. Inventory could be tracked via an online, virtual system. Telecommunications costs could be slashed as participating firms seek other interested parties and link them into the network to get the most favorable tariffs. This bottom area is the one most applicable for e-commerce techniques. It is here that we consider network affiliations, marketplace buying, consortium pooling of category purchasing, and so forth. It is an area for the e-commerce-adept buying professional and a cadre of trusted supply chain partners to reach out for the highest level of mutual benefit.

As mentioned, electronic procurement will eventually eliminate most of the tactical operations in any sourcing function, so it becomes an inevitable part of the supply chain progression. Seventy percent of subscribers to *Purchasing* magazine indicate they are already actively pursuing e-procurement. In many firms indirect material sourcing is well on its way to becoming an automated process. *Information Week* reports that 36% of suppliers are connected in electronic versions of supply chain, based on a study of 500 companies. So the transition is well on its way. The questions become where is it headed and what are the benefits?

Getting to the Extra Benefits Becomes the New Game

As a firm and its sourcing group determine a move should be made toward e-procurement techniques and some of the advanced relations with suppliers mentioned, there must be an inducement for both parties to participate. Continuation of the relationship and commitments to volume are only the entry fee at this stage. Doing business over the Internet always sounds good as potential partners consider getting to the next level of savings, but it has many inhibitors.

General Electric Company provides an example of the inhibition. GE announced through its Chairman, Jack Welch, that a cyber-based move would be made by that firm as he told the business world that GE expected to save $10 billion in cost through its Internet technology efforts by the middle of 2002. The company has now scaled back that estimate as only $1.6 billion had been recorded by mid-year 2001. Now no firm is going to turn up its nose at $1.6 billion, but the GE move shows a company has to be careful with its up-front expectations and be prepared for the inevitable pushback.

One of the reasons given for the scaleback in expected savings is that GE is experiencing difficulties connecting customers and suppliers to Web-based trading systems. One analysis splits GE's supply base into 7,500 already working over the Web to perform functions like order processing and logistics management, 7,500 connected via EDI systems, and 15,000 relying on manual processing (Moozakis, 2001, p. 2). AMR Research analyst John Fontanella has predicted that GE will only be successful with about 60% of the supply base. That is a sobering estimate for companies looking at a higher level of sourcing improvement. It would appear most suppliers are not ready for or willing to make the transition.

GE is not backing off its stance to conduct most of its business over the Web. It is simply recognizing a reality of advanced processing between network constituents. Some are ready and some are not. Some like the idea and some do not. Some prefer to sustain the old methods and wait until forced to cooperate. This type of dilemma is faced directly by sourcing professionals as they move to the advanced stages being advocated. Many of the potential suppliers, which would benefit from participating in such a transition, simply are not convinced there is something of value from moving to an electronic format.

The job of convincing those on the fence to cooperate falls to the sourcing group. They must first segment their supply base into those firms that can do a good job of electronic interchange and those that cannot. Next, they do another sort by those with which they want to move electronically because of the positive effect on the buying firm and those to be included later. The final move is to the core group with which the initial pilot programs are conducted and where the parties can develop a sensible list of benefits for the supplying organizations as well as the buyer.

At the top of that list will be the ability to cut transaction costs. Our supply chain database shows that over half of the companies still conduct business via the mail, telephone, and facsimile machine. Cost for handling an order varies by industry, but a reasonable estimate is $50 to $150 per order. When a firm moves to EDI, that cost can be cut to $10 to $15 per order. The cost of installing the system becomes an inhibitor here, but is generally overcome as the speed and lower costs are considered as good payback features. When the move is made to a full electronic system, errors go away, reconciliation is dismissed, and the cost of order processing sinks to pennies. A firm can multiply its number of orders per month by the difference in costs and generally show a very favorable payback. The problem with this part of the analysis is that the savings are from people's time processing the orders. Unless there is a headcount reduction, the savings are hard to track. Most firms reassign these people to other, hopefully more rewarding jobs.

Fortunately, there are other benefits. With some form of e-procurement system where volume is aggregated across the full firm and possibly with other partners in the network, a supplying firm has the potential to build new revenue with very little selling expense or overhead costs. That is an additional inducement. Longer-term commitments for such volume also increase seller interest. The real benefits to be derived, however, come from removing inefficiencies from the supply chain. As firms get their hands on data relating to the actual processing that takes place, they can bring focus to the delays, poor forecasts, extra handling and shipments, emergency shipments, returns, and special price changes that occur.

Firms often find there is a trove of opportunities here to clean up many long-standing problems by using the data to get to the root causes and eliminate them. Now a supplier has less to fear from being forced to give lower prices for higher volumes and can concentrate on saving money for both parties. If the ultimate purpose of supply chain management is to optimize inter-enterprise processing across the full global network, then working together to eliminate the problems in those systems lies at the heart of the effort.

Electronic Buying and Marketplaces are Part of Future Sourcing Efforts

Exhibit 11.3 shows three reasonably simple steps to the progression we are considering. In Phase 1, a sourcing group will consider e-procurement techniques as an advantage given the limited time and resources to do a thorough job on all buying categories. A typical effort involves calling in one or more of the "buying robots" to help. These are firms offering software and systems that automate the selection and buying of many indirect categories. Ariba, Aspect, Harbinger, Commerce One, and others are very capable of bringing early stages of automation to what is typically a very mundane part of purchasing procedure. Working with such firms can get a firm started on e-procurement and lead to further supply chain efforts. Ariba has now launched Ariba Enterprise Sourcing, software that adds automated contract management, supplier negotiation, and statistical analysis to auctions and online procurement systems. Aspect is now aligned with i2 Technologies and can combine constraint elimination with a suite of supply chain software offerings.

Small teams or individuals will use such software as they try to find better buying conditions in this phase, without having to do all of the arduous

Phase 1 — e-procurement techniques	Phase 2 — network collaboration	Phase 3 — full network connectivity
Consider moving some categories to an electronic format; test buying "robots" for nondirect materials and service Use selected auctions to find better pricing and features in direct areas Apply techniques to reduce order execution	Enlarge the participants in the supply chain improvements discussions Focus on nonprice issues and problems; develop lists of mutual improvement opportunities Conduct diagnostic sessions to pinpoint process step improvements; build core of key supplier/partners	Consider communication needs across the total network; evaluate the disparate systems and agree on the extended enterprise platform Build online, virtual systems to view and track raw materials, WIP, and finished goods; automate ordering and billing

Exhibit 11.3 A Framework for advanced sourcing.

searching through catalogs and interviews. Others will progress to using selective auctions where they seek to buy commodities or some direct materials. Our experience indicates some very interesting first- or second-attempt efforts in this area bear fruit. One study in the United Kingdom found "average savings for materials purchased through business-to-business auctions amounted to 17 percent across 6 categories of goods" (Kambil, 2001, p. 53). Certainly, the use of the robots and auctions will reduce the transaction costs involved.

In the second phase, attention moves to the firm's position in an end-to-end network and the total costs in that network. Now the sourcing professionals enlarge the number and functions present in supplier discussions aimed at lowering these total costs, time cycles, and customer-satisfying systems. Consideration transfers from what the lowest price is to how we can improve the overall supply chain. The focus goes to nonprice issues, problems, and opportunities as teams are mobilized to develop lists of actions that could enhance earnings for all constituents — buyers and sellers. Some form of diagnostic exercise is generally conducted in this phase to get at these opportunities, with strong recommendations from the sourcing function influencing which firms participate. By analyzing each step in the linked processes, teams typically find plenty that can be improved.

In the third phase, a very special group is formed to design and develop full network connectivity. This effort is for the most advanced of supply chain networks and may be in the future for the average players. Now the partners consider the information flows across the extended enterprise and fairly evaluate all of the disparate (but currently functioning) software and systems in use by the various constituents. An architecture design team is usually formed to sort out the better and best features and decide on the platform for use across the network. The result will be an online, virtual communication system that allows participants to view in real time such things as: where the materials are, where the work-in-process is, and where the finished goods are. Matters of order entry, order processing, inventory management, and payment are automated features of such connectivity. In this phase the sourcing function has been helpful in building network relationships that elude normal arrangements.

Concurrently with Phases 2 and 3, the advanced firm will consider some form of electronic marketplace, or trading exchange, that enhances network collaboration and precedes a fully automated communication system. That means most firms will test one or more forms of a Web site where buyers and sellers meet to carry out business transactions quickly and efficiently, without the usual paperwork. Unlike EDI, which requires special software at each end of the transaction, using the Internet means that any browser device can be used to access the exchange. Several options appear for the would-be practitioner.

A firm could begin by contacting other companies interested in consolidated sourcing. A nucleus firm could get in touch with a group of suppliers, other firms of equal size and interest in different industries, or a combination of suppliers and customers in their own industry. The purpose becomes to discuss how certain categories of purchases can be aggregated and put in the hands of one member of the association to buy for all members. Working in areas like telecommunications, office supplies, MRO supplies, transportation, information technology, and travel, such arrangements can lead to better pricing. As the volumes increase, most of these efforts are turned over to a third party entity to handle the transactions, measure the savings, and control the temptation by some maverick buyers to circumvent the negotiated contracts.

A company can also align itself with other companies in a public marketplace consortium. Combined industry experience is used to work with an existing supply base, hopefully to get better pricing and special features regarding holding inventory, finding lower shipping costs, and moving to electronic payment. VerticalNet is an example of this type of consideration as that organization has created exchange communities for more than 50 different vertical industries. Each VerticalNet exchange caters to very narrow focus groups with such sites as SolidWaste.com and Nurse.com. Using the right exchange will allow the buyer to go beyond the simple buying and selling of direct materials and services to consider a full range of other features that complement the commercial transactions. For example, a buyer for molded parts might go to PlasticsNet.com to find a mold manufacturer, a special plastic resin, and a logistics provider.

As the souring group is moving along the progression suggested in Exhibit 11.3, the most appropriate trading place for a particular firm will become clearer. Of particular interest will be the private trading exchanges. AMR Research predicts these marketplaces will represent as much as 52% of the revenue flowing through B2B processes by 2004. In this situation, a nucleus firm and a few allies can form a buying portal or private exchange and thereby consolidate purchasing power. This option will bring the necessary focus to what has been a very trial-and-error phase of strategic sourcing and should eventually be the prevalent option. Dominated by a few large industry players or one significant nucleus firm, they will survive the testing era and emerge as the e-business force in sourcing. As a resort for those still unsure which avenue to take, a company could turn to a hosted service where sourcing is done electronically by a third party. Free Markets and Ariba Sourcing are options in that regard.

For the present, sourcing managers are still plowing through a myriad of options. The number and nature of online marketplaces continues to grow, producing private and public, vertical and horizontal models, all offering significant opportunities to improve total cost of ownership. Some of these entities survive and increase membership. Others are short-lived and fade from the scene. As firms sort through the options, they generally accept that if they do not participate in this trend, they will risk being left behind as e-procurement techniques bring initial savings and the learning necessary to utilize the site for further process improvements. One study conducted by Roland Berger Automotive Competence Center, for example, indicates that $600 per vehicle through e-business techniques can be taken out of the North American truck industry. Our studies indicate more than $3000 can be taken out of the cost of making and delivering a car.

Buyers also realize the risk with aligning to a venture that lacks staying power. Hospitality marketplace Zoho, which was founded by Starwood, owner of Westin Hotels, and Sheraton lasted less than a year. Ventro Corporation created Chemdex.com to be a resource for chemists and scientists, but had to close this site early in 2001. Ultimately, the trading exchanges that survive will be those that go beyond the idea of being an industry marketplace to offer sophisticated software and services, including supply chain planning and demand planning functions. Our favored choice is an industry marketplace fostered by one or two dominant nucleus firms.

Further improvements to the sourcing process will come from the understanding gained in these early efforts and the eventual electronic linking of buyers, suppliers, marketplaces, and external systems. Such enabling will improve order processing, production planning, storage and shipping, and financial processing. These are areas of supply chain not typically considered a part of the sourcing function, but very much a part of advanced efforts. That is why it has been suggested that an absence of these techniques is another supply chain mistake.

Firms that want to implement these further improvements, however, must be prepared for one final inhibitor — the cost of participation. As advanced B2B technologies are introduced to network processing, a firm must be prepared for investments in the necessary hardware and software, which can amount to 25% of the savings that will be generated after installation. There is also the cost of reorganizing and transforming functions and roles that is often overlooked. It is not an easy process and typically requires about 3 years for full and successful implementation.

Our experience is that marketplace options will continue to demise, contract, merge, and otherwise be reduced to a small number per industry, dominated by a few large players. Of the successful ones, Exostar stands out in aerospace with such major players as Boeing and Lockheed, as does Pantellos with American Electric Power, Cinergy, Duke Energy, El Paso Energy, and 17 other leading North American utility companies.

Sourcing managers are well advised to continue monitoring the options and carefully work on a limited basis with those offering the most promise for a particular firm, possibly a buying portal anchored by a nucleus firm.

Portals initially target indirect spending, but eventually expand to the purchase of strategic goods. Suppliers do tend to resist these portals unless there are clear benefits established for them. Since cost transparency leads to pressure on prices and not the features we are suggesting, these portals should be a part of advanced efforts. With just a few industry leaders driving the portal, work can move to value-adding features and longer-term positions for suppliers. Portals are relatively inexpensive and can deliver almost immediate productivity gains to those who use them. With growing repositories of data being a feature of this form of marketplace, members can track trends, add context to business information, and provide a clearer understanding of business conditions.

Action Study — Global Electronics Company

At the 1999 Purchasing Summit in Florida, the chief purchasing officer (CPO) for a U.K.-based firm told 300 attendees that, "We will present one voice, one vision, one set of expectations to suppliers." Representing facilities that comprised the full global electronics and engineering company, these attendees were gathered to formulate a single purchasing strategy. The company is the result of a merger of two culturally different European firms. Following its amalgamation, the $16 billion firm began operating with four different divisions — Software Systems, Automation Systems, Control Systems, and Powerware.

The CPO's plan was for the firm to optimize values from its supply chain. By consolidating purchasing at the newly merged company, he expected to reduce costs by over $500 million over 3 years. The company spends in excess of $7 billion annually on such goods and services as metals, electronics, chemicals, plastics, and MRO supplies. Responsibility for consolidating sourcing for the business units belongs to the CPO and his team. At the summit, he posed some pertinent questions to his purchasing audience: Do we know how much we spend? With which suppliers? Do we know how many suppliers with whom we do business? Are there purchasing policies or procedures in place? When the answers were not to his liking, the CPO enlisted the help of the IT function to help develop a database to gather information across the diverse firm.

"We quickly created the database," says the CPO, "and we tried to make it easy to populate." In gathering information, he recalled that some companies had sophisticated MRP systems and were capable of readily providing the required data. Others were still working with pencils and paper. Completing the job of gathering and organizing the information took perseverance but, with strong senior management support, they were able to achieve the goal. Purchasing now has 600,000 records in its database. Having the appropriate data strengthens their negotiating position and removes much of the emotion, according to the category manager with responsibility for indirect materials purchasing.

The sourcing group collected data through the company's buyers, including such information as part number, supplier, manufacturer number, volume purchased, and last price paid. For each part sourced by a unit of the company, the purchasing team assigned a Dun & Bradstreet number for easy cross-referencing within the firm. With a compilation of the company's overall spend from each business unit, sourcing determines the largest users of a specific category and the size of the spend. Three times a year, each unit uploads its most recent 12 months of purchases.

Taking a look at one category, the company's springs purchases, the group determined through information gathered from the database that the firm had been doing business with 86 suppliers. More important, they learned they were not purchasing springs from two of the industry's top manufacturers. Initial savings resulted from consolidating the supplier base to six.

Tracking the Results

To arrive at the $500 million target, the CPO and his group measured savings by purchase cost reduction achieved as a result of the new buying procedures. These reductions are those that reach the company's P&L from the period of the merger until the end of the first 3 years of operations. Direct material savings are measured from purchase order (prior to a new agreement) to purchase order (after the new agreement is in place). The tool to measure these savings is the "cost savings tracker." Indirect materials are tracked in a similar manner when possible. Certain commodities, like corporate travel and MRO, are not covered by a purchase order (PO). Many of these commodities are measured by establishing a base discount rate and comparing purchases against the new discounted rate.

The cost savings tracker comprises a central database of contracted savings and actual savings. This information serves as a forecasting tool for plants and divisions to use for budgeting and as a measure of the effectiveness of the organization in achieving the potential savings. One part of the tracker provides the user with contracted savings information; specifically, this is segmented via signed purchase contracts or a firm PO by supplier and commodity. The savings recorded usually are entered by a global category manager. Sometimes, when a commodity is unique to a single plant or two, a unit purchasing agent can enter the effort. That individual uses the same delineated purchasing process as the global manager.

Another part of tracker measures the implementation process. When a plant or division utilizes the contracted savings, the unit enters the difference between the purchase order price prior to the contract against the new price. The difference is entered each time the contract is used for one year after implementation. The tracker is available via the sourcing Web site on the company intranet and is controlled by password. Specific reports of savings are available. Reports are produced for the entire corporation, each division, and each plant, by commodity. All new agreements have an execution plan by a category manager.

The Purchasing Summit

At the purchasing summit, the CPO and his team launched the consolidation effort. The intent was to ensure everyone understood his or her role and to unite sourcing within the firm. Any boundaries that existed between plants, companies, and divisions had to be destroyed. The chief operating officer (COO) was also on hand to lend management support to the group. He quickly made it clear to the attendees that he had a thorough understanding of the sourcing function and its value to the combined company. "Purchasing is an area where we have leverage," he told the audience. "A lot of our competitors do not have this capability. No single element within the company can save this money." Divisional chiefs expressed their support as well.

The summit included working sessions that broke out attendees into groups designated by categories such as metals, plastics, chemicals, electrical equipment, and electrical components. These groups verified data that had previously been collected. In one session, attendees broke out the categories into groupings and assigned a commodity code for each group, identified the code owner, the level of contract ownership, and established code member teams. Another session had attendees identify projected savings over 3 years for each category and assign individuals to implement the savings. More than 190 opportunities for cross-sourcing and cross-selling were identified during the 2 days of the summit. Relevant category teams analyzed and then developed specific action plans.

Summit organizers spread the word throughout the company: an article on the summit with a summary of the initiatives was included in the company's monthly newsletter; a video of highlights was distributed across the firm; and a quarterly global purchasing newspaper continually highlights the initiatives. Through the breakout sessions, the category teams were able to identify the specific expected savings from the $4.7 billion annual buy.

The Sourcing Structure

The structure for global sourcing contains some very interesting features. A supplier development team, which is made up of "black belts," is a crucial part of the effort. These managers help assess future suppliers and work with the business units to implement new sources in their systems. They are responsible for supplier quality audits, product capability reviews, design process improvements, and implementations. Divisional coordinators have also been appointed to help implement global agreements within their divisions.

At the summit, a determination was made of which categories would be purchased by the global teams, the divisions, or by an individual plant. Forty-four global category managers are responsible for developing strategy, helping select suppliers, and negotiating the contract. Business units are responsible for ensuring materials arrive at the plants at the time of need. The goal is Six Sigma quality (less than 33 bad parts per million) delivered on a just-in-time basis. Black belt teams reinforce the process with quality audits and continuous improvement at the plants. Members go through special training and commit to spending up to two years full time running Six Sigma projects.

Responsibilities of global category managers include understanding the customer's business needs, preparing the organization to utilize the global purchasing initiative, data collection and communication, developing the purchasing strategy plan for the category, defining the category, preparing a position statement for the category, evaluating and selecting suppliers, finalizing contracts, capturing cost savings, and expanding the process for further savings or improvements. With the help of the business unit, they define the appropriate communication channel to deal with price changes, multisite delivery/quality issues, and contractual changes. Monthly reviews are conducted to track progress and results.

Global Purchasing Process

One of the tools used is the global purchasing process, which calls for active involvement by the business units. Through this process, the category managers work with a core multidiscipline team of individuals from the business units, including quality, engineering, legal, and finance function representation. The process includes these steps:

- Select category and identify its code. Managers base decisions on data gathered by the business units.
- Download the necessary information from the central database.
- Implement the plan. Category managers are responsible for handling issues that might surround implementation of an agreement.
- Invite relevant users to the initial category meeting.
- Create a position statement for each category published on the company's Web page.
- Present nondisclosure agreements to relevant suppliers.
- Create a musts/wants lists to rate suppliers. Each supplier is graded on these lists and the high score determines the preferred supplier.
- Hold a possible supplier quotation conference or create a list of possible suppliers.
- Generate a mini-bid or full request for quotation (RFQ).
- Possibly audit supplier quality.
- Complete needs/wants list for each supplier.
- Select the supplier. The team favors long-term, world-class suppliers.
- Negotiate the contract. This activity is generally performed solely by the category manager.
- Business unit sign-offs on supplier selection and savings calculations are completed.
- Sign the supply agreement and release for implementation.
- Publish an abbreviated version of the agreement on the Web site.
- Implement agreement; track and report savings; work with suppliers for continuous improvement.

"I cannot stress enough," says the CPO, "the importance of working as one organization. We will minimize costs through a single Invensys purchasing process, establishing lanes of responsibility that are coordinated across the organization. This means no turf wars. We optimize savings, not for departments, but for the company." And one division chief executive added, "The purchasing train is leaving. If you don't want to go where it's going, you had better get off now."

Summary

There will be a continual pressure applied to the sourcing function to bring savings from suppliers to the bottom line of a P&L statement. That is the inherent nature of sourcing as senior executives expect buyers to wring every possible dollar from relationships with suppliers. As supply chains move through the evolution to advanced stages, a more mature and mutually beneficial environment can and should be created. The key is to begin work with a core group of suppliers with whom a long-term position is taken. The above action study illustrates one methodology for getting to that stage. The effort continues as key partners work with nucleus firms to find network savings. In the most advanced stage of this relationship enhancement, the parties work together to automate as much of the processing as possible, using the Internet as the primary means of communication.

12 Mistake 11: Dealing Incorrectly with the Existing Culture

Working with companies that extol the virtues of supply chain and the need to integrate technology and collaborative techniques into the marrow of the effort, I find a surprising paradox. The avowed dedication to moving a firm to the most advanced levels possible comes across early and loudly. Proponents indicate strong endorsement and support for things technical and collaborative. At the same time, the preparedness for the ingrained business cultures to accept and assimilate the necessary transformation process is often lacking. Despite published accounts of one or two success stories (usually in one or two business units within a large firm), the majority of firms studied suffer from this paradox. The heart is willing, but the mind is sluggish.

Most companies are deeply into some type of continuous improvement effort that has been embraced as part of a larger supply chain initiative. Under skilled direction, these efforts move as close to optimization as possible. Most of the involved firms also have a culture that favors incremental change for the better, fueled by internal-only effort, and backed with a lot of denial for why objectives were not achieved or when exposed to evidence of better competitive progress. In supply chain, this condition limits a firm's movement across the supply chain evolution. When a company's culture stands in opposition to a new strategy or better practices, the culture always wins, and we encounter the next obstacle to progress — dealing incorrectly with the existing culture.

This factor can slow or inhibit progress at any point along the journey toward supply chain optimization, but it is critical at the point of transition between Levels 2 and 3, where a firm moves to a network focus and use of external resources or remains fixated with internal excellence. In the third level of the progression, a firm must rely on network partners and use e-commerce capabilities to further enhance any supply chain effort, often in spite of resistance from internal groups.

Firms can effectively work together to improve such supply chain processes as order processing, planning, and logistics without overt collaboration involving many functions. That has been done for years. But none of these accomplishments will be totally effective unless they take advantage of the enhancing properties from the cyber world and the collective learning and experiences that exist across an extended enterprise. Making use of external resources, Internet technology, and cooperation from willing partners demands that those responsible for advancing a supply chain deal effectively with the culture so it does not impede this progress. Such action begins with a hard look at the capabilities and cultural inhibitions within the company.

Transitions to the Cyber World Face Internal Limitations

Collaboration and use of the Internet technology have introduced an entirely new means of conducting major portions of business transactions. With these tools, advocates found a new set of opportunities to enhance functions and actions in the end-to-end processing that takes place across a supply chain. They discovered the ability to simultaneously involve many functions and experts from willing partners in the pursuit of optimized process steps. They also found a cadre of challenges to good implementation, beginning with pushback from nonadvocates — those who really do not understand what the whole transition is about or how external help and cyber technology can dramatically improve their area of responsibility.

How well firms can develop an e-commerce-enabled supply chain network that takes advantage of capabilities of most or all of the partners in the extended enterprise will be a key factor in which companies succeed or fail in the future. With limited success stories to guide the transition, however, this challenge looms as the next hurdle for most aspiring supply chain leaders. Among the limited case studies, one does stand out as an example of how a firm can come to grips with its internal culture. Boeing delivered a world-class example of successful online collaboration when the firm designed its new 777 airplane in virtual cyberspace.

By electronically sharing design tools and processing techniques with engineers, customers, maintenance personnel, project managers, and key suppliers of components and subassemblies around the world, Boeing, acting as a nucleus firm with its network partners, redefined how airplanes are built. According to company sources, there were no paper blueprints. The work was done interactively over the Web. A new slogan captures the spirit of the effort — "the 777 is a bunch of parts flying in formation." Boeing's customers no longer have to wait 3 years for delivery of their orders, as the firm now believes it can deliver in 8 to 12 months. Boeing also believes it has increased its capacity to build more than double the number of airplanes in a year.

This type of success story is what wins over the nonadvocates, but it does not assure completion of the task. The journey to advanced supply chain management begins with an understanding that, in spite of feigned interest and support for things technological, many people lack the technical skills to take advantage of these new tools. When asked by senior managers about what the largest barriers to moving from Level 2 to Level 3 are, my first response is that it is a matter of people getting away from self-interest (in career, function, and local urgencies) and moving to a true customer and network orientation. Directly behind that problem is obstacle number two: the lack of understanding of what Internet technology is all about, how it applies to enhancing a business strategy, and what meaning it has for individual employees. In spite of all the rhetoric on the importance of adopting cyber-based technology and collaborating with key partners, the real understanding of the value and means to exploit these values resides in only a few hands.

A study conducted by the Economist Intelligence Unit (EIU) and Meritus Consulting, LLC validates this observation among business managers. In this study, when senior executives were queried on the importance of the Internet, 82% said they "believed Internet technology would have a major impact — or even totally transform — their supply chain performance." At the same time, when these same individuals were asked about their preparedness for using such a tool, "the vast majority acknowledged that they were ill prepared to integrate this technology into their business processes."

One example helps describe the problem being considered. Most manufacturers or processing firms expect their suppliers to deliver precisely what they want at the point of need at a 96 or higher rate of efficiency. At the same time, few organizations reach that level with their customers. The same study reported that only 20% of respondents said their firm achieved such a level with customers, and 50% acknowledged they lacked the functional or management expertise to make e-commerce a competitive weapon. From a general perspective, 35% said their company lacked "the technology to implement a digitally enabled supply chain" (Ljungdahl, 2000, p. 83).

Now every businessperson knows what the Internet is and how it is sweeping across the B2B2C landscape. What they do not understand is how deeply it will impact virtually every step in supply chain processing. The real e-business revolution is about transforming the way business is conducted. Many of the participants, however, fail to understand the proper techniques for enhancing manual process steps (often fraught with mistakes) and developing fail-safe, automated procedures that bring a new level of productivity and financial results to a firm. When it comes to introducing cyber-based solutions across a network, this lack of understanding is a real obstacle to overcome.

There is simply an army of people in business who have a half-understanding of what the new information technology tools are all about and how they can be used effectively with external partners. In the absence of better understanding, these people often turn to software that purports to greatly enhance business processing, rather than redesigning and building better systems that can then be automated with the appropriate software. The results can be very problematical in those instances.

The challenge is not isolated to one part of the world. Following the last British elections, the Labour Government declared that its goal was to establish the U.K. as the best place for e-commerce by 2002. This was a lofty and well-intentioned declaration. However, one study, called *The Quiet Revolution*, released in February 2001 by the Confederation of British Industry (CBI) and KPMG, reported "over 76 percent of U.K. companies currently generate less than 5 percent of their turnover from Internet economies." A follow-up study conducted by British consultancy firm Rubus, reviewing the *London Times* top 1000 companies, showed 90% of respondents acknowledged "the importance of an effective online strategy to their success." Only 15%, however, had "developed such a strategy for 2001" and only half had a "dedicated head of e-business" (Walton, 2001, p. 24). Our global experiences verify similar conditions in most countries.

Transitioning an organization from a Level 2 position, where supply chains have been improved, but primarily for the good of specific functions or business units, to a real external orientation and focus on customers and consumers, takes more than a rallying cry from leaders. It requires working effectively with the existing culture and advancing through five steps:

1. Raising the awareness of the general body of a firm to supply chain concepts and the need for technology and collaboration as tools to move to advanced levels of progress
2. Winning over business unit leaders as sponsors, so significant members of the firm can take a leading position in the advanced movement
3. Understanding the hidden values in supply chain and e-commerce enhancements
4. Recognizing the strengths and weaknesses of technological solutions
5. Moving forward with tests and pilots that prove the validity of the concepts

Raising Awareness across the Firm Positively Affects the Culture

To combat the absence of full cross-enterprise understanding of the values inherent in a supply chain improvement effort, it would seem logical that special education would be a solution. Unfortunately, there is a real absence of solid advanced supply chain education today. Although many organizations and institutions offer supply chain courses, most are of a sourcing, logistical, and internal planning (or Levels 1 and 2) improvement nature. And the would-be supply chain professional attends most of these courses. That does little to help the cultural inhibitions being considered.

There are a few exceptions to this contention as some corporations have established supply chain training as part of their curriculum in their internal universities. This training is generally all internally focused, however, and rarely includes any external parties or work of a collaborative nature. Most courses are also offered on a sign-up basis, which means business units can disregard the training or send a very limited number of attendees.

The first step in overcoming cultural inhibitions to advanced supply chain efforts is to create educational courses and insist on participation so the general body of the firm can raise its awareness of what supply chain is all about, what it can mean to company performance, and the disadvantages of not participating in the full evolution. So long as the general knowledge across a firm is limited to improving sourcing and logistics, as most course work today considers, efforts are going to be restricted to an internal excellence focus. Moving to network formation includes external partnerships. That means eventually the training has to be extended to cross-organizational participation.

Technical Skills Are a Premium Commodity

As a firm does move to Level 3, it begins work of an external nature with a few of its closest allies. In this area of the evolution, the company works with network partners to gain an advantage from pooled knowledge. Among the issues considered will be:

- How to collaborate to grow the size of the overall market
- How to compete aggressively to gain market share in this larger market
- How to strategically outsource core process steps to the firm having the greatest competency
- How to increase shareholder value for all members of the supply chain network

Answers to these queries come from sharing of best solutions among the partners. The value of the answers, however, is a function of the skills and experiences of the practitioners involved. The requirement is to bring together firms and people who have a credible awareness of what Level 3 and beyond means to the company and its network partners. In the absence of advanced training and skills, the collaboration will be weak. Consider the need to integrate disparate ERP systems, a key tool of advanced supply chain efforts. Culturally, this will be an enormous challenge as each constituent will have limited trained resources and will want to favor its selected ERP system. Collaboration demands these constituents work out the means to integrate the valuable information flows from their systems across the extended enterprise. Today that is being done on an ad-hoc basis.

There is also a major cultural inhibitor in this area. Common wisdom has held for some time that enterprise resource planning would not fit e-business modeling. In the first place, much of the ERP software was based on old technology and very internally focused. Second, supply chain partners would be loathe to share valuable internal information with external parties. Now that the Internet offers the potential to extend planning (and the matching of demand signals with supply capacities) to suppliers and customers, a new advantage has appeared, and ERP can be at the center of e-business models. Therefore, it must be a part of advanced supply chain training, and we have yet to see a training course offering this education on a cross-enterprise basis.

Some firms move ahead in spite of this limitation. Anne Chen, writing for *PC Week*, notes that, "Companies such as Turner Industries, Owens Corning, Super Discount, and Crestone International LLC have spent millions of dollars deploying ERP, and are now building on their initial investments and making it a critical component of their e-business strategies" (Chen, 2001). Chen also advises that the goal is to link ERP systems to the Web and open traditionally closed systems to partners and customers. Supply chain partners in such a system can now work collaboratively to provide end customers with real-time information from their linked data systems via Web browsers. But such a move takes a lot of skilled people and learning that does not currently exist. Waiting for success stories to initiate action is a slow process. We favor education and inspiration. That requires another (external) step in the training process.

Four Steps to Better Training

To meet the need being considered, a four-step process is advocated:

1. Organize in the list an internal training effort, which focuses on the fundamentals of supply chain theory and practice and how beneficial results can accrue to the firm. Skilled and proficient trainers should be processed through the formation of this effort and the specifics of the course instruction and materials. That means train the trainers. Let them be a part of creating the course syllabus. With their help, search out the best applications inside and outside the industry. Build a supply chain framework that fits the organization and matches the needs of the business plan. That requires consensus on what the characteristics of each level of progress will be for the firm. Anticipate the cultural pushback. Consult with key business leaders for advice on how to make the education meaningful to all parts of the firm. Get agreement on the suggested characteristics. Let the trainers develop the course work and training aids.
2. Begin with a cross section of the organization and conduct a series of training sessions intended to raise the awareness of the entire firm on the advantages of moving through a supply chain evolution. As feedback is received, modify the course so it reflects what is needed to overcome any and all cultural inhibitions and to take advantage of the most current best practices. Keep the material current by accessing best practice information and action studies from around the world. Expand the training as more business units sign up members. Encourage these business units to conduct their own training across the unit to get the understanding as deep as possible.

3. Expand the training selectively with a small group of key partners from the extended enterprise. That means select a few allies interested in the same improvement to awareness and understanding, which share a stake in the success of mutual efforts. Now the training brings together the best thinking of several parties dedicated to optimizing a total supply chain effort. Again, the trainers would be involved first to determine how the course material can be integrated and reflect best practices across the network. The training takes on a new dimension here as the material must cover how to integrate information flows vital to optimized processing between the parties. This work should be applied to a test group representing some of the more experienced people from the collaborating firms. With critique and advice from the test group, the course should then be offered to internal personnel and attendees from the involved firms.
4. Offer appropriate portions of the training to interested suppliers, distributors, and customers from the second and third tiers in the supply chain network. This technique assures important players will be up to speed as the firm decides to expand its supply chain effort and move all units into the advanced levels. A nucleus firm is best positioned to augment this phase of the training with a group of key supply chain constituents.

Education is not an easy proposition to sell, but it is essential if firms are to continue to reap the benefits of supply chain efforts. Moving to the advanced levels is a very slow process as a study of many industries is proving. Closing the gap with the leaders and getting the most from the effort requires a higher level of awareness than exists today. Understanding the true benefits of technology and collaboration as beneficial tools is clearly another need. A solid training format will only increase the possibilities of meeting these needs and dealing with a firm's culture so it enhances, rather than inhibits, progress. With that format in place, we proceed to the next step — securing the support of the business unit leaders.

Winning the Minds and Hearts of Business Unit Leaders Is Crucial

One discovery about the success of supply chain efforts stands out over all other factors I have considered. It is the realization that forceful and dedicated business unit leaders must take their part of the firm to Level 3 and beyond before the organization as a whole will consider such a move. A strong CEO can influence a firm to make a commitment to advanced progress, but few sustain their interest long enough to see a large company make such a journey. Functional leaders can offer strong support, but their efforts must be accepted by the business units to be successful.

Better understanding of the benefits of an advanced supply chain effort will excite members of the organization, but may not be enough to create the momentum to move to the advanced levels. Most often, a single business unit or two will prove the value of external collaboration and the importance of building a network approach to supply chain so others can follow. With documented evidence that sharing information and linking Internet technology systems together has extra value, more business units will emulate these leaders.

If the road to success must be taken one step at a time, and cultural barriers will be the most difficult obstacles along the way, then the best advice is to select a business unit with a strong, dynamic leader and an ethic for innovation and need to be ahead of others in an industry to lead the movement. This translates into choosing a visionary leader responsible for a business unit of importance, has a good industry reputation, and has achieved a reasonably good record for process improvement and financial results as the initial Level 3 sponsor. The leader should also be a person who is reasonably adept at Internet technology and possesses the internal desire to prove what can be accomplished with collaboration and the application of that technology. That is the person to lead a firm into Level 3 and beyond, and convince the nonadvocates to come along.

Since this leader will be closely scrutinized by the entire organization (and often criticized by the nonadvocates) during the eventual testing and pilots that take place, the CEO must be involved in the selection and issue the necessary support for what will be a time of difficult transformation. Care should also be taken to understand just what the unit does not know about collaboration and technology. Selecting a great leader with a poorly equipped technical group and a history of alienation between buyers and sellers will doom any cyber experiment. The trainers in the suggested awareness format would be well advised to conduct the first internal training session with such a leader and his or her key people.

Remember the Need for Skilled Talent

Leadership and dedication are not the only necessary ingredients. The presence of skilled talent is what determines the level of success with the effort.

In the study previously cited by *The Quiet Revolution*, it was discovered that the majority of British business managers are optimistic about the future of e-commerce. While this is not a surprising finding, the report went on to say the lack of progress with e-commerce initiatives is related to "the initial lack of both the correct skill sets needed to maximize the commercial benefits and understanding the significance of this new method of doing business" (Walton, 2001, p. 25). That translates into not having enough skilled technicians and supply chain designers who can use cyber technology and collaboration to enhance a firm's performance.

Once the selection of the appropriate leader and business unit have been made, the best talent from across the company must be made available to help in structuring what becomes a design experiment having importance for everyone in the company. Results from such efforts, unfortunately, indicate progress is slow, so the commitment for special resources — from sourcing, operations, information technology, and logistics — will be hard to attain. Once again, CEO oversight is mandatory if success is to be achieved. Approaching the effort as a test that will have importance for the entire organization is the typical mandate applied for securing at least part-time commitment of the necessary experts. Indicating that the culture must adapt to documented improvements also helps getting the right resources for implementation.

A Supply Chain Model Guides Advanced Efforts

The process now moves to designing the extended enterprise that will be enhanced through collaboration and technology. Using the business unit as the nucleus firm, drawing on the training received, working with key partners on both sides of the unit, and applying the techniques described in earlier chapters, the unit designs a new e-business model. This model must have at its core strategies based on collaboration and technology that continue to reduce costs, make better use of collective assets, build new revenues, and bring a level of satisfaction to the targeted customers not currently achieved in the industry.

The model developers must begin by drawing a depiction of the end-to-end process steps that describe the way the test unit conducts its business. Numerous supply chain models are available as guides in this regard. Attention should be given to capturing the product, information, and financial flows across this depiction. Throughout the process, care must be taken to indicate where critical information sharing should take place so an online condition can be created for network partners. That is where the skilled technicians earn their merit badges.

The Internet has provided business with a powerful tool that can be used effectively for process improvement and growth opportunities. It is a new way of conducting business that is all about transforming the way a business operates. Better costs, better-managed assets, increased revenues, and higher customer satisfaction can only lead to greater profits — and the Web can enhance all of these factors. But using the new cyber-based tools effectively can be the ultimate challenge to a firm, particularly one comfortable with using traditional, labor-intensive systems. For a firm lacking Internet skills, it can be a showstopper.

Using a liberal amount of external advice, the modelers now establish a list of improvement initiatives that can enhance the test unit and move it to a world-class condition. Since the model will become the device that secures endorsement and gains the support of other business unit leaders, a special step must be taken.

Describing the Model Requires Good Articulation

As the model is developed and the necessary action steps defined, the test unit is prepared to begin pilot operations, but everyone will be watching progress. That means the model designers will have to explain how it will be used and how the results will be tracked. That requirement leads to the next consideration — how to make certain the existing culture does not sidetrack or de-rail the effort. Advocates should make certain those less adept at technology and collaboration see the value of the process changes and the payback for what will be a sizable investment.

It means winning over the less informed through explanations, demonstrations, and presentations of results in a manner that does not turn off the less technically skilled members of the audience. That requires a skill beyond technical expertise. It requires spokespeople for the test unit who are articulate at explaining just what is happening. If these people are not present within the selected business unit, they must be imported, even temporarily, to explain the validity of the advanced levels of effort.

Raising Awareness of Hidden Values Secures Support

With general education accomplished, the leader selected and business unit concluded, a design model that makes sense to viewers in place, and the availability of sufficient technical talent secured, the firm is ready for the next step — raising the awareness of what the hidden values are for all parts of

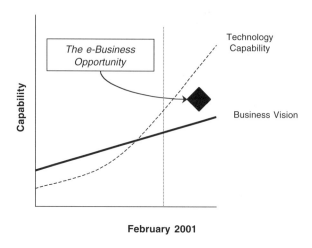

Exhibit 12.1 Grasping the e-business opportunity.

the organization. As much as the supply chain professional may advocate the importance of advanced practices, firms move when employees across the company see meaningful benchmarks or specific results that prove the value of the effort. As most of these values are unknown, it requires a dedicated group digging out metrics and information that document the difference between current performance and what could be achieved. Previous chapters have listed many of the potential savings. Now the task is to document this potential in terms meaningful to management and employees and have the articulate spokespeople deliver the message.

Exhibit 12.1 illustrates what is being pursued. Any firm is going along an existing path defined by the business vision and supporting plan. All such paths lead to better performance. The e-business opportunity, which builds an increase to the slope of this performance, raises the curve dramatically through technology and collaboration. Describing and measuring that opportunity is what explains the hidden values from advanced supply chain efforts.

To move a company to the Level 3 digitally enabled supply chain, there must be a group deployed to evaluate all of the advanced tools and measure the results of application. Among the many options, my favorites include

- E-commerce, e-business, and emerging technology tools, including Internet, intranet, extranet, artificial intelligence, Groupware, wireless communications systems, Web-enabled knowledge management, supplier relationship management (SRM), and customer relationship management (CRM)

Mistake 11: Dealing Incorrectly with the Existing Culture ■ 185

- Enterprise resource planning (ERP) tools, including those offered and presently implemented by SAP, BaaN, Oracle, Peoplesoft, JD Edwards, and others
- Advanced planning and control systems, including those offered and already implemented by Manugistics, i2 Technologies, and others
- Advanced collaborative planning, forecasting, and replenishment (CPFR) initiatives

These tools should be studied for potential application for added value or a determination made of current results if already in place. If partial achievements have been made, the analysts must go beyond those metrics to determine what the full impact can be. The degree to which individual constituent performance, and eventually shareholder value, can be improved needs to be documented. That requires an honest and frank assessment that is often best conducted with a combination of internal and external resources. A framework often helps this assessment.

Exhibit 12.2 uses one measure of performance, economic value added (EVA), as a barometer for measuring improvement. Other measures, such as

Economic Value Added (EVA) =

Net Operating Profit − [Cost of Capital × Capital Employed]

Sales − Cost of Sales   Net Working Capital + Net Fixed Assets

Sales   Current − Non-
                                    Assets   Interest-
                                               Bearing
                                               Current
                                               Liabilities

Material + Labor + Overhead   Premises   Plant
                                    &   &
                                   Land   Machinery

Impacts of Successful Supply Chain Management

Exhibit 12.2 Creating higher results through supply chain.

return on net assets employed (RONAE), return on investment (ROI), or some evaluator of return on effort, can be used. The results will be the same. The model depicted shows EVA as a function of net operating profit, cost of capital, and capital employed. Beneath the typical formula are listed the individual elements of EVA and how they must change to gain an improvement to the overall measure. With this exhibit as a guide, the team must determine where collaboration across the supply chain, using technology tools, will increase or decrease the appropriate measure. In conjunction with the process map defining the end-to-end supply chain for the business unit, the analysts can determine:

- What will be required to design the "virtual business model"
- The need for accessing information to identify the value creating opportunities
- The basis for designing the data flows to enhance an e-business supply chain model
- How to form a "value sharing" scheme to tap the best practices of all constituents in the network

To get a firm to adopt e-commerce for performance enhancement and use the listed tools collaboratively, companies must reevaluate their position in an industry and market, find what further improvements are feasible, determine just what capabilities are missing, and what is needed to change the organizational mindset to take a leadership position. They must then make an assessment of how the test unit can enhance its industry position through answers developed from the above determination.

That starts with understanding the changes occurring in an industry and how interactive, digital technologies are having a positive impact. It also requires firms to take a more holistic look at the extended enterprises and develop e-commerce solutions that ensure meeting strategic objectives. This means firms and their closest allies evaluate e-commerce in terms of how it will help transform business processing and not just the technical infrastructure.

Companies are aware they must make some type of transition to digital technology to remain competitive. Few have the deep understanding or metrics to prove the case or make the requisite investment. To counter such conditions, use the business unit as its test case and proving ground for how to move the full organization into the digital world. With few information architectures up to the task of taking the unit to Level 3, the testing and piloting becomes a time when the entire company can benefit by keeping the risk isolated to one part of the firm. Now the combined best technical resources use their collective thinking to analyze the business conditions mentioned and construct an e-business model for optimizing performance.

When we deal with such a business unit in the pilot stage, the first question to be answered is how much does the senior management team really understand about the benefits of an e-commerce-enabled supply chain. Our experience indicates the answers vary by industry, from a low in construction, forest products, and industrial equipment to a high in aerospace, telecommunications, and high technology equipment. In-between are all of the other industries slowly making their way forward with what can be a gut-wrenching transformation process.

The analysts have to calibrate the test unit and honestly establish the magnitude of improvement that is possible, given the skill set with which the unit must work. The process then moves forward in this area as the business leader and his or her key staff come together with company specialists and selected external advisors determined to be necessary in an e-business model-building exercise. Start with a good analysis of the gaps between the unit's performance and best in industry class, or best in any industry class.

Next, the participants establish a best-case scenario, which depicts conditions in Levels 4 and 5 of the supply chain evolution. Drawing on the analysis and case studies considered earlier, these future state conditions are generally an eye-opener to observers. Our experience with many firms indicates that first-pass presentations show so much improvement that observers are overwhelmed. Improvements that include 50 to 60% reductions in cycle times, 20 to 30% reductions in inventory needs, and 5 to 10 new percentage points to profits often appear to be unbelievable. The test group is advised to build its new condition format in stages. That means build the savings into a 6-month, 1-year, and 2- or more-year outline that depicts the improvements as they accrue to the unit.

Understanding Technical Solutions Enables the Transition

With the potential benefits displayed, the test unit now turns its attention to helping participants understand how technical solutions will be an enabling factor in transitioning to the proposed new state conditions. The requirement that makes technical solutions a viable part of daily activity in an organization is an integrated information architecture. This requires significant involvement of the CIO and his or her staff members, much investment, and considerable patience as disparate equipment and systems are harmonized into successful network connectivity. Real-time, highly accurate information transfer must be at the heart of such connectivity. One of the most serious mistakes supply chain practitioners make is assuming the existing data connections in an extended enterprise supply chain are capable for conversion to e-commerce interaction across the network. Adding a digital data transmission capability across multiple network partners to an existing, but flawed, infrastructure only worsens the results. The nonadvocates will rally around this condition.

By frankly assessing the capability of current systems, with the help of the IT department, supply chain partners, and knowledgeable experts, firms can establish the requirements and measurements that assure an e-commerce-enabled supply network will provide a competitive advantage. For example, the ability to forecast demand and plan appropriate supply chain responses is an absolutely essential element of advanced network activities. Such a condition requires access to historical data by more than one party, online tracking of actual performance by all constituents, and information availability in the proper format for each network member. This is an area where participation by willing partners will definitely help. Collaborating with a supplier or two, a key distributor, and a key customer, a group can find a wealth of improvement ideas. As each of these constituents shares experiences and successes with various technical solutions, the unit gets much closer to the best solution.

Those working to move the test unit forward should also select action cases from within the industry or solicit the help of software vendors to visit companies having successfully applied such technical solutions. Demonstrations are a useful first step, but should be augmented with conversations with actual practitioners willing to explain the benefits and problems associated with moving to what usually becomes automated processing.

There is so much software being introduced in supply chain that a thorough evaluation is well beyond the scope of this text. A team from the test unit should consider the framework being used and decide where the most critical points of information need occur, where manual processing slows the cycle times, where online visibility will have the most advantage, and so forth. With the help of collaborating partners, a prioritized checklist of areas that can be enhanced with technical solutions can be prepared and the team can search for successful implementations. Armed with positive results, they can dramatically increase the firm's appreciation for how these solutions can improve performance. Then the transition from existing methodology to an e-commerce orientation can be enabled.

Test and Pilots Prove the Validity of the Concepts

The firm will make progress by next establishing a test or pilot with the selected business unit. For a predetermined time frame of 90 days or more, this unit should test the proposed technical solutions. Using the recommended process map with prioritized improvement initiatives, or the EVA or other model to calibrate and measure improvement, ideas derived from the collaboration and technology assessment are applied.

There is no roadmap for this part of the exercise. The few documented case studies, such as Boeing discussed earlier in the chapter, have resulted from inspired leadership and ad hoc processing. A firm simply has to make a commitment to test and validate the best ideas and innovative concepts that seem to apply or have applied in other situations or industries. Working from a prioritized list or potential implementations, the unit works through a series of experiments with new software, systems, and collaborative adoption of value enhancing tools.

Mistakes will be made as often as progress is recorded. Both experiences are valuable learning for the overall firm and should be carefully documented and explained. From the pilot will come the evidence that either convinces the firm to move forward or cements the culture in place. My experience shows the former situation occurs much more often than the latter. With success validated by the test unit effort, the next step is the hardest. The effort must be extended across the balance of the firm and best practices not only documented but also made a part of normal business processing. This is the work of the CEO and the senior executives of the firm.

Summary

All firms are moving forward with some form of improvement effort. Most are merging these efforts in an end-to-end supply chain initiative. How far the firm progresses with that effort is highly dependent on the existing culture, which will dictate whether the effort is restricted to an internal focus or that external resources will be used. Moving to an external environment also requires the adoption of advanced techniques, including cyber-based technology and network collaboration. These features are generally poorly understood in most organizations and require a concerted effort to change mindset and dedication or the culture will derail the effort.

This chapter has presented a five-step procedure for raising awareness of the values inherent in advanced supply chain management through education and the testing of potential improvement techniques with a designated business unit. Gaining the support of management and employees in a firm will be a function of how well these designed experiments are conducted and the positive impact of the results. With confirmation of the best applications and documentation of what does not work, the culture generally accepts the mandate to move forward.

13 Mistake 12: Not Trusting the People You Need to Trust

Take it from an expert. Frank Quinn, former Chief Editor of *Supply Chain Management Review*, has captured the new thinking. He avows, "Forward thinking companies now realize that how effectively they collaborate with their trading partners — both up and down the supply chain — will determine their overall business success. The ability to collaborate with your trading partners has become a core business competency." He quickly adds, "Without question, technology is a critical component of collaboration" (Quinn, 2000, p. 222).

Quinn is speaking to both the supply chain professional and the general business manager when he advocates getting on the supply chain technology train and riding it with your best trading partners to the next level of process improvement. To his advocacy, we add an important caveat — without the participation of trusted partners, it will not happen. At best, technology is the key enabler of advanced supply chain efforts and collaboration sends advocates in search of the best applications across an extended enterprise effort. These ingredients are the new and most powerful tools available for business process improvement. Together they offer great promise for the future. But without partners who trust each other and share what is necessary to apply the tools, it is a fruitless endeavor.

A large nucleus firm usually drives most Level 3 and higher efforts, so it should be expected that trust would be discussed but not necessarily given appropriate consideration as supply chain alliances are formed. That is another mistake. In this chapter, we will take a final look at what can stop a dedicated effort to get to the highest possible level of the supply chain evolution. We will consider the last obstacle to success — not trusting the partners you need to trust.

Smoke Gets in the Eyes Before Sensible Propositions Develop

It is a fact of supply chain life. Companies get started on a supply chain effort and make considerable gains as they focus on their internal needs. As the results diminish, they come to an important realization: that alliances with external partners will be necessary to reach higher levels of accomplishment. Now they send emissaries in search of willing partners to assist in further steps in the journey. Since no single firm has all the skills to get to the highest level of success alone, it makes perfect sense to elicit this help.

There is another less publicized fact of supply chain life — most of the alliances fail. It is not because the intentions were not good. In the beginning, all would-be partnerships look good to the proponents and generally have a well-founded charter. Most are, in fact, properly conceived and full of potential opportunity. What invariably destroys the potential is actions by members from both sides of the alliances, some of whom never wanted the partnership in the first place and who do what they can to disrupt the processing. My learned analysis of watching and working through many of these situations is that the element of trust was not sufficiently discussed in the beginning of the negotiations or thoroughly understood across the enterprises before the partnering pacts were signed.

Consider the example of Concert, the joint venture between AT&T and British Telecommunications. The idea was to provide large business customers with global telecommunications services — a noble concept. The alliance made good sense as one viewed the proposed structure. As a single entity, neither firm had a full global capability. Rather than invest the huge capital necessary to achieve such capability, why not establish a jointly funded entity that would? The entity was formed, followed by considerable publicity favoring the alliance, and an enormous struggle to make it work. Concert now loses about $200 million per quarter and the internal bickering has been viewed in public.

This example is but one of many where the parties got carried away with the concept before they had worked out the processing that would have to take place and the need for people throughout the collective organizations to endorse the new entity. Clearly, a novice could have seen that it would take the merging of two different cultures — British and American. It would also have required determining how the partners needed to depend on each other to carry out their part of the bargain across a domain as large as the full globe. From my viewpoint, the level of trust necessary for such a venture to work was never present.

There is a fundamental problem here. These types of alliances are a necessary part of advanced supply chain management. It is that simple. A firm is not going to get to Levels 4 and 5 without the help of partners on both the upstream and downstream side of its supply chain. Amazon.com is playing this scenario out in front of our eyes as that company continues to make alliances with firms crucial to sustaining their existence. Several have been referenced already, but one makes the point. Toysrus.com could not make a go of e-marketing and selling on its own. Amazon.com had to increase its offering beyond books and tapes. Making money together through the electronic channel made sense and seems to be working for both firms.

At the same time, other arrangements do not seem to get off the ground. Covisint and Transora struggle to make sense of their partnerships with multiple automobile and consumer product firms, respectively. These organizations are trying to get very powerful firms to cooperate so obvious benefits can be achieved. After several years, the struggle continues as the proponents seem unable to get the parties to take advantage of these exchanges. It is trust, people. Can't you see it?

Most partnerships fail because at least one party (and often both) tries to control the relationship. That negates the definition of partnership. It certainly does not match the meaning of collaboration. When this happens, all attention goes away from what could have been achieved and passes to: how do we handle the political ramifications? We then find a condition cited by veteran business observer James Champy. He comments, "Both parties just spend too much time musing about the fantasy of possible synergies and too little time figuring out what it really takes to make an alliance work" (Champy, 2001, p. 23). As mentioned, nucleus firms appear in my crystal ball as the eventual leaders of higher level supply chain collaborations. Making those collaborations work requires such large and often very proud cultures to accept the fact that control is not as important as the value of the cooperation that could result.

There may be a textbook soon on how to effectively form supply chain alliances, but it does not exist today. Documentation needs to be developed on proven alliances that stand the test of time. In the meantime, formation of a successful network alliance is a mandate in search of the necessary ingredients, many of which have been covered in previous chapters of this book. My first-hand experience tells me trust sits at the center of those ingredients, but gets quickly overwhelmed by concerns for control. To get through network formation and on to Levels 4 and 5 of the progression, firms must learn to set aside self-interest. They need to work on the beneficial aspects to be derived from mutual processing that could take place across an extended enterprise. Smoke has to be removed from the eyes, problems must be anticipated, and sensible propositions have to lead the partners to the success they envisioned.

The Problems Can Be Internal and External

It is an internal and external problem. From an internal aspect, companies launch new supply chain ventures without the necessary long-term understanding of what is involved and commitment to the people who will dedicate their energy to the effort. Kmart launched an electronic selling channel cleverly called BlueLight.com. After initial successes, the site seemed to miss the ambitious targets set for it, and the firm announced plans to spin off the site and reintegrate the service into its stores. Establishing an entirely separate business for well-known brands has always been a difficult effort, so the designers should have been aware of potential problems. There is no way to know exactly what commitment was made, but departing BlueLight.com executives have been outspoken in their dissatisfaction with the parent company. This is not a situation to blame on the Internet or the dot-com crash in general. It is a situation that needed more trust, a planned time frame, contingency alternatives, and a reasonable exit strategy if objectives were not met.

Internet Week analyst Tom Smith cites a similar situation at giant retailer Wal-Mart when he notes, "Dot-com doomsayers will latch onto Wal-Mart's plans to reintegrate its spun-off Web business as more proof that the Internet was never more than a technology fad." Smith asserts that such a conclusion would only demonstrate the folly in this thinking. "The best and brightest managers selected by Wal-Mart," he reports, "were hampered from the outset because they couldn't fully exploit all that their parent had to offer: strong supply chain relationships; a mature, first-rate technology infrastructure; premier branding; and retailing expertise. It's fair to say the integration wasn't adequate" (Smith, 2001, p. 1). One could also wonder if the die-hard, traditional retailers within the firm failed to help the new cyber group try to introduce a new way for consumers to get Wal-Mart products.

Trust begins within an organization. After a lengthy business career, I have concluded it is either there or it is absent. Working with numerous companies to build a viable supply chain effort, the first challenge is to take advantage of best practices and synergies across the organization. The few firms with the requisite level of trust seem to move easily in that direction. Sharing of expertise and resources occurs and improvements are spread across the entire firm. Those without it never do tap the full potential of the organization. They hoard their resources and constrict information on best practices from each other. These firms are content to let individual business units progress on their own. That is a formula for suboptimization and petty politics.

When trust is absent on the inside, there is little chance it will work on the outside. Suppliers can be invited to participate in share sessions, for example, but these quickly degenerate into a means to save money just for the buyer. Only when there is an up-front agreement to share the benefits do these exercises continue to progress. We have worked with enough groups where there was a trusting relationship to know it can work. As supply chains mature, it is time for those responsible to come to grips with this necessity and begin to build the external pilots that prove you can trust some people and firms, especially when there is a mutual interest. It is also time for CEOs to reevaluate their internal cultures to determine just how much trust there is on the inside of the firm, before parties are sent externally in search of supply chain allies.

Key Enablers Are Crucial to Success

Supply chain professionals are well aware of the necessity to include trust in collaboration, but now there is evidence to support that contention. A study conducted by John T. Mentzer, Professor of Logistics at the University of Tennessee, with sponsorship from FedEx Corporation and *Supply Chain Management Review*, has identified the enablers, impediments, and benefits of supply chain collaboration. Based on feedback from 20 supply chain professionals representing a broad range of industries and business sectors, Frank Quinn reports, "The bottom line of this research is that supply chain collaboration can deliver a host of benefits *if the right enablers are in place* and if the obstacles (both internal and external) can be overcome. Somewhat surprising, these enablers have more to do with management style and interpersonal relationships than with technology" (Quinn, 2000, p. 222).

Examples of the enablers included the following:

- Common interest and clear expectations. All parties need to have a stake in the collaboration's outcome to ensure long-term commit-

ment. All parties need to understand what is expected of them and others in the relationship.
- Openness and trust. For a relationship to work, the partners must openly discuss their practices and processes. Sometimes this means sharing information traditionally considered proprietary.
- Recognizing who and what are important. Not all partners and supply chain activities are created equal. Choose the ones that will deliver the greatest benefits.
- Leadership. Without a champion to move collaboration forward, nothing significant will ever be accomplished.
- Cooperation, not punishment. When things go wrong in a relationship, punitive actions seldom make them better. The right approach is to jointly solve the problem.
- Benefit sharing. In a true relationship, the partners need to share both the pains and the gain.

While there is not much here that might not have been intuitive, the survey confirms the necessary ingredients for moving forward. And the thinking represents more than just supply chain advocates — a number of supplier-customer pairs participated in the study to make sure that aspect was included. The participants were united in their belief that the key enablers are crucial to supply chain success; they concluded that, if an organization and its partners make certain the enablers are in place, success can be achieved and the benefits of collaboration realized. At the center of that certainty will be the trust imperative, a commodity missing in most advanced efforts.

Before moving to that imperative, let us consider one organization and how it has progressed with a key supply chain constituent. Goodyear has been a fixture in the tire business for a long time, and moves a significant amount of product through a solid dealer network. Back in 1997, the firm decided it was critical to the future to conduct as much of its business through this dealer network as possible via an extranet created for that purpose. Dubbed XPLOR, Goodyear has been slowly introducing the site to dealers who are traditionally averse to things technical and have limited IT capabilities. About 65% of the firm's dealers, or about 3,000 locations, now make use of the extranet site. The site delivers about 200,000 pages of content to dealers monthly.

The XPLOR extranet allows the dealers to manage orders, receive the latest marketing information, and check prices, inventory levels, and order status. The site also delivers messaging, online collaboration and marketing materials, and product service bulletins. It is an example of how a nucleus firm can introduce beneficial technology solutions in a collaborative manner without insisting on controlling the situation. According to analyst Richard Karpinski, "Goodyear has chosen the carrot, rather than the stick, in moving dealers onto the extranet. In most cases, it's the dealer's option to move business on to the site" (Karpinski, 2001, p. 2).

The firm is constantly searching for new features that will enhance the site, as recently it began to manage dealer rebates via the extranet. Goodyear is also looking at adding the ability for dealers to process invoices and payments online. Yet to appear is better use of sales information to boost efficiency through the extranet. As a leader in proving the validity of an electronic B2B channel with distributors, the company worked with the dealers to make the site friendly and applicable. Duane Hand, Goodyear's e-commerce manager, explains, "It's important to make sure dealers don't view the B2B portal as a 'back-room-type solution.' We want to try to get it to the front counter where it can become an integral part of a dealer's operations" (Karpinski, 2001, p. 3).

That kind of thinking is what is necessary up front of any move to an electronic enhancement. It takes a lot of working together to find what the real advantages will be, how they will be used, and how the user can transition from a manual state to the new cyber world. Goodyear also set up a help desk to walk dealers through XPLOR installation and training, and using feedback to assure the new system was as user friendly as possible.

The Trust Imperative Is Crucial to Collaboration

As companies around the world begin to embrace collaboration as a necessary tool for advanced supply chain management, they make a discovery. The advantages to be gained through external cooperation require the sharing of information previously considered proprietary and not suitable for external eyes. Data on sales consumption (which relates directly to current demand), production schedules (which reveal pending supply), product design (which informs all parties of what is forthcoming), and logistics (which explains where everything is across the extended enterprise) must be viewed in order for a Level 4 and 5 progression to exist. These are the elements of attaining a competitive advantage for the network. Trusting that the supply chain partners will act responsibly when they gain access to that information is not only crucial to the success of the collaboration, it separates those who succeed from those who fail.

We are not talking ethical behavior here. If that is not present in the firm, nothing can be trusted and you have picked the wrong partner. We are considering professionalism. Collaborating effectively is only possible in a business setting when the parties establish expectations in the beginning and abide by the defining charter during implementation. What data do we need, how will it be gathered and shared, who should have access, and what simply will not be made available are questions to be answered in the first meeting. Then it becomes a matter of doing what the partners said they would do (that is the external need) and assuring the folks back home that they will also abide by the agreement (that is the internal need).

Dixon Ticonderoga is a large manufacturer of pencils, a firm that also has a proprietary chemical mixture used in the manufacture of soybean-based crayons. We would never expect that firm to give access to that formula any more than we would expect Coca-Cola to share its carbonated soft drink formula with external parties. That is something that must be made clear. Sharing important data among partners goes beyond such issues and moves to what is important to reduce supply chain costs across the total linkage, how to make best use of all assets, how to draw new revenues through the network, and to keep the business customers and consumers happier than they would be dealing with another network. As soon as collaboration is considered as a means of favorably impacting these factors, the data needs should be defined, access determined, and processes set in place to facilitate proper use. Anything beyond the determined usage should be declared off limits and honored by the professionals in the agreement.

Sometimes the need for cooperative behavior extends to things considered outside the realm of alliance partnering. Procter & Gamble, a firm dedicated to enhancing customer relationships, for example, decided that creating a more collaborative environment within the firm required a change to their time-honored system of compensating brand managers. Sales quotas were eliminated and business development teams created with key customers, starting with its flagship partner, Wal-Mart Stores, Inc. Product managers began to receive compensation for the success of the entire supply chain effort, not just how many cases were moved. As an example of how the effort has paid off with better data from its customers, "today, P&G uses collaborative forecasting for 45 percent of its U.S. sales and 33 percent of the products it sells internationally" (Wilder and Soat, 2001, p. 38).

Put the parameters on a piece of paper at the first meeting. Give all parties a chance to critique the list at the meeting and with important internal resources back home. Once the list is refined and covers how the collaboration will proceed, sign an agreement that firms the charter and spells out what is in the deal and what is not. If patented processes are out of bounds, then make that data off limits to the users. Above all else, do not try to conquer the entire universe in one fell swoop. That means do not start with a collaborative scope that is too hard to monitor and control under any circumstance. Begin with pilots and the data needed to enhance relationships in one small area. As success is generated, expand the charter and include a larger area of focus. That is how you move to Level 5.

To illustrate how a firm moves toward the levels of trust needed, let us consider the experience of retailing giant The Limited, as that firm succeeded in building collaborative technology solutions for the benefit of the entire enterprise. This company has developed an e-commerce strategy focused on building real-time connectivity between the IT group and the databases in the back office with functions necessary to operate at the store level. The engine for this effort became a wholly owned subsidiary, Limited Technology Services (LTS). It started internally, grew with help from the business units, made a few mistakes along the way, and persevered on its way to success. As the firm learned to build trust internally, they moved with pilots to external partners.

At the center of the initial strategy were focuses on using the Internet to automate outbound logistics management and helping the operating units, which includes more than 2,700 stores and such brands as Express, Lerner New York, Lane Bryant, and Henri Bendel. Through Intimate Brands, of which the firm owns an 84% share, another 2,460 stores are represented, including Bath & Body Works, Victoria's Secret, and White Barn Candle. The Limited is a far-flung enterprise with business units serving specific consumer groups and possessing unique supply chain requirements.

In 1995, the firm entered into a transformation process with a growth objective to turn each retail business unit into major individual brands. Using IT as an enabler was at the heart of that goal. At the outset advocates knew the existing IT strategy had to be changed. As president Jon Ricker explains, that strategy was intended "to clone its original mainframe environments and run each business as a separate silo from a technology perspective. The challenge was that we had individual retail brands that were not big enough to support the cost of a high-quality, large-scale IT organization" (Karpinski, 2001, p. 2). With these thoughts in mind, the firm set about to design uniform systems that could positively impact supply chain processing and win over the support of the individual business unit leaders.

Since the thinking at the time favored IT support systems that were partially centralized and partially decentralized, the original scheme was to centralize oversight of these operations but allow a degree of local control as well. What evolved was LTS, established in August 1999, which operates as a centralized technology and consulting company and has its own P&L plan. Ricker calls it a virtual technology company that can put any resource against any project.

The unit started by consolidating 14 data operation centers into two, one domestic operation in Columbus, OH and one global supporting operation in Andover, MA — not an easy task for a firm that had grown dependent on multiple storage and delivery sites. The requirement, of course, was to convince the business unit customers that there would be no decrease in service.

In an effort to consolidate technology as well, the firm made dramatic moves in eliminating much of its legacy systems to bring a uniform design to the effort. LTS brought apparel brands onto one format and retail management onto another. A common warehouse system was also installed, backed with a data warehouse and reporting platform. "We went from something like ERP to something we call GRP, global resource planning," Ricker explains. "We went to a best-of-breed strategy and integrated it deeply at a data level."

With the help of vendor Tibco Software, the firm handled the need for middleware integration and built an enterprise portal to process the data. The idea was to move from batch processing to a real-time, online information platform. That feature was intended to help the unit managers and store mangers see where the inventory was and move it to the point of need. It took a lot of work and cooperation across the enterprise as everyone had to adjust to a new way of doing business, but with a degree of freedom at the brand level. Three years into the effort, The Limited is now seeing payback at the business unit level, thanks, in part, to the value of the insights provided by operating managers.

Ricker and his team continue to fine-tune the effort. They are now moving into supply chain automation, working with 45 of the firm's largest trading partners to build a private logistics e-hub that will help automate the distribution side of the business. "Improved Web links with suppliers and e-manufacturing efforts will come later," Ricker expects. Assisting the firm is Logistics.com, which now operates a Web service that allows The Limited to move communications with shippers to its Web site. The retailer has "improved its average delivery time by a week and cut shipping costs by more than $1 million, or 3 percent." Next steps include improving the data flows between LTS and the individual stores (Karpinski, 2001, p. 3).

Suppliers and Buyers Must Be Ready and Willing to Make Collaboration Work

When a firm decides it is at a point where an excursion can be made into network formation and has at least a small group of willing partners interested in helping, the issue of trust must be part of the initial discussion

Mistake 12: Not Trusting the People You Need to Trust ■ 201

	Thought	Plan	Action
Design	Collaborate electronically on design/development projects — new products, models, prototypes, or components	Match CAD-CAM capabilities with design needs; integrate systems, conduct preliminary tests	Establish pilot operation with one to two new efforts, from easy to hard; track progress, modify system, and report results
Source	Automate sourcing processes with select group of suppliers; consider VMI features	Consider consortia leverage; form pilot groups; identify initial categories; select software; determine metrics and how to share savings	Set time frame for test to prove concept; begin first phase category sourcing; measure savings; determine fee structure
Plan	Integrate ERP systems to share demand and supply information	Select areas where data will be shared; settle on how to transmit info across different systems; build a model for using information	Exchange demand planning and capacity data on specific products in a test mode
Make	Transfer part of manufacturing or full contracting to core competent partner	Reduce manufacturing costs, shorten cycle time by moving process steps to best-able partner	Select process steps, parts, subassemblies for selective outsourcing; monitor results
Store	Use shared warehouse with network partners; manage inventory with online visibility	Create system to track end-to-end inventory; develop model to match warehouse needs with customer demands	Rationalize warehouse system; build pilot for network virtual inventory and storage system
Ship	Use virtual logistics system to reduce transportation costs	Transfer ownership of shipping assets, driver liabilities to 3PL or lead logistics provider	Establish test sector, select logistics software, determine partners, and begin pilot

Exhibit 13.1 Preparedness chart.

and made a part of the eventual processing. A form of strategic supply chain due diligence is necessary if collaboration is really going to work. That means the parties must come together and define the parameters within which they will collaborate. They should define, for example, the information environment they want and the steps to be taken to assure both parties and their organizations abide by the accepted parameters.

Exhibit 13.1 is a simple guide that can be used in such initial meetings. The parties to the collaboration should develop their own guide, but this exhibit provides a framework for beginning. Along the vertical axis, the steps in a typical supply chain process are listed. Across the horizontal axis, we list three steps that will take the participants from **thought** to **plan** to **action**. Now the grid is filled out together as the partners spell out how they will proceed together.

In the design area, for example, the group starts by describing the thoughts they have in mind that could improve new developments. The exhibit shows "collaborate electronically on design/development projects" and lists a few examples. We proceed then into the plan section and decide to "match CAD-CAM capabilities with design needs and integrate systems through preliminary tests." Then we move into an action stage where we suggest establishing a pilot operation with one or two new developments, from easy to hard (to test the ability to collaborate at different degrees of importance), and track the process so it can be modified.

Each step should be considered in whatever the group decides are the end-to-end processes in their network. If selling should be included, for example, then that category should be listed on the left of the grid. The point is to have a living document that spells out just what the parties are planning to do as they collaborate and share technology. This grid should then be reviewed with other functions and managers at both firms so there is an upfront understanding of what is going to take place, and decisions can be made on the necessary resources to achieve implementation.

Many firms will recognize this procedure as something that used to be called value discovery. The difference is that such efforts used to be limited to representatives from buying and selling, with an occasional designer or engineer. The new format calls for IT specialists, operations personnel, planners, engineers, logistics experts, and so forth to get as much vertical interface as makes sense to build innovative solutions to traditional supply chain processing steps. The internal reviews of the working documents will then be crucial to determine the level of organizational readiness to proceed. A key ingredient here will be determining the ability of the firm to hold its own as the collaboration goes forward with partners that might be at a higher level of development.

Not all companies are as ready as others to enter Level 3 and higher of the supply chain progression. Perusing the grid and discovering what type of commitment will be made also forces the company to evaluate its capability to be an effective partner. Some of the skills that will be required include:

- The degree to which the firm's IT organization has the time and skills to provide expertise on hardware, software, and programming needs.
- The ability and willingness for members of the sourcing function to cooperate and offer nonpricing suggestions to the dialogue that must take place.
- The level of the firm's sophistication regarding e-procurement, e-commerce, e-logistics, and so forth, and its ability to collaborate across disparate operating systems.
- The level of progress with the firm's supply chain improvement effort. It is hard to match a Level 2 and Level 4 business unit in a meaningful collaboration.
- The readiness of the firm's communication infrastructure to share meaningful information with a selected group of external partners.
- The level of understanding and support for such collaboration at the most senior levels of the firm, including the degree of commitment to make valuable resources available for team exercises.

This list can be expanded as the participants review the readiness grid and determine the actual implementation steps that make sense to get to the potential benefits. Its importance lies in helping the companies decide just how ready both sides of the collaboration are to participate in what becomes a well-scrutinized effort, one that will determine whether the firms goes forward with more collaboration or sinks back into a focus on internal excellence.

Action Study — Sun Microsystem

Sun Microsystems is a $16 billion company with offices in 170 countries, a firm providing end-to-end technology solutions ranging from desktop workstations to supercomputers. Sun, as a nucleus firm, decided some time ago to move beyond outsourcing of logistics and transportation to collaborate with key service providers to mange information and improve processing.

In one particular move, the firm decided to connect with 12 principal suppliers for parts distribution and repair in the Americas. After a decade of work with this initiative, the firm reports it has tightened performance against customer requirements from "two days to minutes." This effort offers an excellent example of how patience, testing, and collaboration can work.

Sun began in the early 1990s to outsource various aspects of its service and repair needs to companies they believed were skilled as repair operators, transportation providers, and warehouse management firms. The moves were made to allow Sun to benefit from supplier capabilities and expertise and to concentrate more adeptly on core competencies. An early discovery was that coordination was a real challenge. Every time a part moved from one party to another, a transaction occurred that had to be recorded, monitored, and audited. With nearly 150,000 demands for parts per quarter, the participants knew they were dealing with too much redundancy, chances for error, and detailed information that was used by multiple partners.

Sun's response was to move boldly forward and integrate its technology platform with those of the 12 key suppliers. In the process, 60% of the former touch points were eliminated; two distribution centers were consolidated into one; and the firm moved to an exception-based management system. Since each of the external suppliers had a different operating system, Sun played the role of a nucleus firm and worked with its partners to develop the integrated platform that was eventually used. Equally bold targets were set as the firm decided to shoot for a 50 to 80% improvement in effectiveness every 2 to 3 years.

In 1994, Sun established its Americas Enterprise Services (AES) as a service arm to handle installation, integration, repair, and maintenance of its systems. The unit operates with 10,000 employees and has annual revenues over $2 billion. Every quarter AES receives 150,000 demands for parts. The unit began by consolidating 70 vendors into 25 and identifying 12 as key suppliers. The firm shares strategy, direction, and intent with this core group and eventually set them up as an *Alliance Council*. Senior-level managers at these companies have committed to an integrated partnership with Sun.

In 1995 the partners introduced e-mail exchanges, not a small effort at the time. Getting everyone used to using the Internet was a bigger challenge than expected. Next, they moved to flat file sharing, then to jointly developing meaningful applications that benefited both parties and, eventually, to using completely integrated applications. Today, the partners operate with a common platform and rely on one major system so they have a seamless set of process interactions among all constituents. Of importance, all parties are on this platform "with the same level of capabilities," according to Peter

Pazmany, senior director of Enterprise Services (Pazmany, 2000, p. 57). To keep current, as applications are developed and used they are continually assessed against business requirements and modified as required.

An important feature demonstrates the role of each supplier/partner in the new environment. When the firm reduced the distribution centers to one site, the warehouse management supplier was called on to progress from being in charge of warehousing to providing critical information. It was a difficult transition for that partner, but today this firm is a key element for the network, supplying advanced information technology and the advance notices of movements throughout the system. The effect has spread across the network. Because parts no longer sit idly in distribution centers, suppliers have increased flexibility to schedule work and smooth out the typical peaks and valleys in ordering that occurs. These suppliers now function more as manufacturing organizations than quick-response troubleshooters, according to Pazmany.

Success with this early effort has allowed Sun to form Sun Logistics Virtual Network (SLVN), integrating systems across all suppliers to execute exception-based management. SLVN links all providers of warehousing, regional stocking, transportation, and repairs so Sun can look at the entire supply chain for returned and repaired parts. Third-party logistics (3PL) providers use information from this online visibility to identify problems that previously would have been detected much later in the processing. Under the old system, if a part were late, it would be detected after the fact, perhaps 2 days later. Now, the integrated system brings Sun into the supplier's world to spot the problem immediately.

The system operates much like an air traffic control mechanism. In a room with few people and loads of monitors, items are displayed in a color-coded fashion (green, yellow, and red) to pinpoint those that are delayed. When an item moves to red, the owner of that item is notified. Green items are not touched. With this system, the debilitating details confounding logistics are eliminated and the supply chain results become more predictable. Because the firm now manages information instead of parts, business decisions are improved and the firm moves closer to an optimized state.

Other process steps were automated as the system developed more sophistication. Field engineers now arrange to send parts, for example, directly to repair vendors, omitting 60% of the touch points and thousands of transactions and its 3PL partners. In one year, the new approach averted 750,000 transactions. Sun determined the system paid for itself in less than 12 months.

As part of a fail-safe technique, at every step in the processing, the group and its advisors challenged the need for outsourcing the procedure. The definition of core competency was applied — areas where Sun made focused, internal investments to maintain or gain competitive advantage — to separate the supply chain continuum into activities that were considered "Sun-centric" or "partner-centric." Then decisions could be made on what stayed and what was outsourced. Internal audit and financial personnel were involved in these steps to assure fiduciary responsibilities were also met.

Regarding the issue of trust, the firm came to grips with the requirements at an early stage. Traditionally, the firm had built IT systems internally, and considering another firm's IT engine was hard to accept. Working with the Sun IT department and other internal information users, AES determined how far it could proceed with information sharing. Some processes were defined as sacred — central to Sun's identity and success. Others were considered more fluid and could be shared with selected suppliers. Exhibit 13.2 is Sun's list of what became known as the sacred and the fluid.

"Sacred" business processes or system interfaces are to be maintained within Sun; "fluid" processes and systems are to be maintained by Sun's business partners.

As a supply chain service partner, AES defines one of its customer metrics as first-time material availability whenever it is needed based on contract requirements. Through the system described, AES has raised that service level metric by six points. A two-third reduction in time to replenishment has

Sacred	Fluid
Purchase order commitments	Detailed transaction levels (including movement between stocking sites and suppliers)
Accounts payable expenditures — cash outs	Field aging of materials
Planning to acquire or deploy assets	Parts call assignment
	Reporting delivery
Financial valuation	Web front-end interface
Item master information	All processes to manage basic inventory flows and movements
Financial book of record	
Asset validation of service	Data feeds to Sun on all basic inventory flows and movements
Physical inventory/cycle count	
Distribution center processing	
Disaster recovery for failed partner	

Exhibit 13.2 Business processes and system interfaces: the sacred and the fluid.

significantly reduced inventory needs and cut millions from carrying costs. There is substantially less paperwork in the new systems. Open-ended purchase orders are used to cover a period of time and reconciled based on tracking flow of parts through the system. When a part is received, it becomes payable. With real-time aging reports, the partners know what parts have been shipped and when.

Summary

Supply chain management only moves to the advanced levels with the help of external partners. Selecting those allies and moving forward in an atmosphere designed for mutual benefit require a level of trust not normally found in such alliances. In this chapter we have called attention to that need and explained a methodology for gaining sufficient trust to proceed.

14 Conclusions: The Path Forward

Throughout this text, a central concept has been fostered regarding a supply chain management effort. The purpose of that effort must be to concentrate on business customer (in a B2B network) or consumer (in a B2C network) expectations, so the system of response can meet or exceed what becomes the manifestations of perceived needs, while making a reasonable profit on the effort expended by all network members. To put that idea in simpler terms, the members of a supply chain network must identify the end consumers (or customers) for its linked process steps and try to determine what exactly those consumers need. With that determination, the network partners work together to optimize the required process steps and satisfy the intended consumers. When done effectively, all parties in the network make more money.

This is not an easy task. Often the consumer does not exactly know what is needed and can be swayed by instinctive impulses. Business customers might be less susceptible to this problem, but are equally as likely to not fully understand what will satisfy a current need. At other times, the consumer has a reasonably clear idea of what to buy but can be diverted by attractive alternatives. A business customer may already have a solution to its understood need, but wants to find other features that reduce total costs. In either event, the linked partners must determine the context of the actual needs and bring forward meaningful solutions. Working together, with data accumulated on a targeted group, allies in a network can generally develop a reasonably accurate picture of what is really needed, what will sell, and what will not.

With that information, the network constituents then draw a process map of what takes place across an end-to-end supply chain that connects raw materials to final consumption. Working in an atmosphere of trust and applying collaboration and the application of enabling information technology, the map is analyzed, dissected, and improved at each link in the processing. As progress is achieved, the consumer needs can be met or exceeded, assets can be better utilized, costs reduced close to optimum levels, and new revenues drawn from the targeted consumers.

Each of the preceding chapters has called attention to the obstacles that stand in the way of achieving this desired state. The 12 largest impediments have been discussed and solutions recommended for eliminating the inherent mistakes. In this final chapter, we turn our attention to how to move forward, given the recommendations that have raised the awareness of what it takes to satisfy consumers and reach the highest levels of the supply chain evolution. Drawing on the recommendations made in previous chapters, we will consider some concluding remarks and the roadmap for those firms interested in achieving the cited purpose.

Start by Calibrating the Firm's Position and Intention

Once there is a reasonable amount of senior management understanding of the advantages offered by higher levels of supply chain progress, and there has been a sufficient amount of training and early level progress, the firm should take time to peg its position on the five-level evolution. By analyzing where the firm stands against characteristics exhibited by other companies in the same or dissimilar industries, the firm can establish its position and determine the advantages of moving forward. This step requires the firm to identify the characteristics of the major functions at each level of progress so an approximation can be made of the gap between current performance and the higher-level firms. Then a reasonable estimate can be made of the value for moving to a higher level or to the desired ultimate level for the firm.

Exhibit 12.1 is an abbreviated form of a calibration chart that could serve as a guide in this area. The progression has been condensed into four levels, with an explanation of the characteristics a firm will exhibit for various functions as it moves from an internal focus to advanced supply chain techniques and on to e-commerce and full network connectivity. A company can modify this chart for specific industry characteristics or use it to help establish the potential for improvement. The idea is to calibrate a firm by function and then develop an order of magnitude of the gap between current-level

Conclusions: The Path Forward ■ 211

Business application	Internal optimization/supply chain optimization	Advanced supply chain planning (some external optimization)	e–Commerce	e-Business
Design and development of products and/or services	Products and services are developed using internal resources only; engineering driven, focus is on best product design; quality is often put in via excessive rework after product or services are produced	Some selected outside resources, such as industry seminars, functional workshops, consumer clinics, or contract design firms, are used.	A team, including selected suppliers, develops products; designs are shared via a CAD-CAM network	A truly collaborative endeavor — each constituent does what it does best and all participants bring a view of the marketplace; consumers are a key part of the design team and a collaborative environment is established
Purchasing, procurement, and sourcing	Volumes of purchases are aggregated at a business-unit level and discounts are sought leveraging purchasing volume	Purchases are understood and leveraged at a corporate level; sometimes suppliers/customers are included to raise volume level for larger discounts	Supplier expertise is sought out and Web-based purchasing is used, sometimes combined with electronic catalogs and/or online auctions; e-procurement and portals are used for some purchases	Best constituent — letting the most capable member of your value chain purchase goods or services for the entire value chain; full electronic (Web-based) catalog available to all value chain members

Exhibit 14.1 Supply chain calibration.

Business application	Internal optimization/supply chain optimization	Advanced supply chain planning (some external optimization)	e-Commerce	e-Business
Marketing, sales, and customer service	Internally developed programs and promotions — account ownership as a sales strategy	Customer-focused database initiatives, marketing clinics, and strong knowledge of customer buying patterns with your firm; advanced telemarketing systems and boiler rooms	Collaborative development of solution sets for customers; joint marketing with key suppliers and utilization of supplier insights; unsolicited fax and e-mail campaigns	Marketing and sales plans developed collaboratively across the value chain focusing on the end consumer; implementation of a consumer response system across the value chain; processes extend across the value chain
Engineering, planning, scheduling, and manufacturing	These functions are discrete without formal linkages and collaboration; tools such as MRP may be used, but the focus is primarily on a single function; multiple processes and/or organizations can originate change requests	Sharing of scheduling for manufacturing is done; engineering specifications are shared, sometimes electronically; first tier suppliers know manufacturing plans for new products; suppliers often use (or are forced to use) assembler tools	Team-focused approach to engineering and production planning; suppliers have visibility and access to CAD/CIM and PDM tools and information; companies have visibility into the supply chain to include customer orders	Collaboration across the entire value chain: customers, distributors, suppliers, and other key constituents; market insight and analysis are shared; best constituent takes the lead on planning, scheduling, design, and so forth; all companies have full

Logistics	Logistics functions are primarily in-house; inventory drives a "pull" strategy (driven by incentives for manufacturing based on high utilization of manufacturing capacity), which rewards a "full truck" even at the expense of delayed shipping for customer benefit	Still often driven by manufacturing capacity, the beginning of a pull strategy emerges; linkages (manual, phone, fax) in place to link orders to shipments (not necessarily end orders — orders to the next stage in the supply chain); some digital channel in evidence	Best constituent provider — can be outsourced partially or totally; the company in the supply chain that is best able to perform operations can lead logistics; the dual channel starts to come into the forefront	visibility of entire chain with the ability to see what changes will impact the entire chain; processes extend across the value chain

There is complete network visibility, including consumers or end customers; more and more content is digital; the build-to-order model is in the forefront for manufacturing; logistics planning is done centrally and consists of in-bound, intra-, and outbound processes extend across the value chain |

Exhibit 14.1 (continued) Supply chain calibration.

Business application	Internal optimization/supply chain optimization	Advanced supply chain planning (some external optimization)	e~Commerce	e-Business
Customer care	Complaint reaction; statistical information is kept on customer complaints, product, or service; updates are done on those complaints with the greatest volume; rebates and incentives are often used to promote sales	Customer care is primarily delivered by call centers; call centers are a mechanism to provide some levels of service to customers seeking information or redress; call centers are primarily cost centers inside a company; some attempts made at new sales (Telco's) to	Primarily the same as the advanced supply chain planning characteristics except help is often available on the company's Web site; also, tremendous advances have been made in auto responding e-mail systems to answer questions; again, many of these activities are in support of cost reduction; companies have been formed to perform remediation and returns for e~businesses; in fact, the creation of another industry segment has occurred	Customer Care will transition into a profit center; customer care activities will work across the entire value chain, with at times a single contact being responsible for the entire value chain; customers will have full access to a customized customer care experience delivered via individualized systems and people; full access to a customer's records and transactions will be available to all value chain participants for support (privileged information is exception); processes extend across the value chain

Human resources	Screening and regulatory compliance; internal functional views for staffing and resources; little involvement in business plans, strategies	In addition to internal optimization, responsible for new work models and seeding academic programs with business requirements; one big U.S. issue was telecommuting as employee life-styles changed	offset expense; product complexity and attempted improvements in efficiency (staff reductions by another name) shift customers to call centers from sales forces, and inside sales forces	Work streams extended across the enterprise; serious supply chain training in place, with external constituents	All resource loading and acquisition will be done across the value chain; processes extend across the value chain

Exhibit 14.1 (continued) Supply chain calibration.

Business application	Internal optimization/supply chain optimization	Advanced supply chain planning (some external optimization)	e-Commerce	e-Business
Information technology	Point solutions and data silos are the norm; no consistent use of technology resources; companies often outsource with little or no strategic benefit at this stage; no corporate technology standards in place, divisions and/or business units set own directions; architecture is normally product based; if there is a corporate architecture, it is from a centralized team and is often an academic exercise that is not used	Corporate architecture is in place and adherence is voluntary but often used; migration from silos is underway and intracompany communications are improved; some published standards are in place to assist in communication with suppliers	Corporate architecture is in place and adherence is mandatory; initiatives are made in establishing intercompany technology standards	Value chain architecture is in place; resources (strategy and implementation) are extended across the value chain; Internet tools and technologies are the dominant technology infrastructure; intercompany is the norm; processes and standards extend across the value chain

Exhibit 14.1 (continued) Supply chain calibration.

performance and the desired future state. With this information, a reasonably good estimation can be made of the value for reaching the highest level that makes sense for the firm.

If, for example, a firm wants to consider the design and development function, it scans the matrix across that category. If the definitions are not sufficient for calibration, they are modified. If an industry perspective is better than the generic statements in the matrix, research should be done within the industry to determine the characteristics of firms at the various levels listed. In the automotive sector, Ford and Toyota stand out as Level 4 and 5 organizations. Information has been documented on how they have progressed design and development of new cars to much shorter cycle times than normally experienced in that industry.

With satisfactory definitions, the firm then pursues the matrix with members of the organization using a cross section of functions, and not just the department personnel, because those members invariably give themselves higher ratings than their colleagues. When there is consensus on the position, members of the function then set about to make a reasonably good estimate of the value for moving to the desired level. We use desired level because it may not be necessary to reach full network connectivity for each function, unless a competitive advantage results from having that level of performance.

With a position calibration for each of the major functions and an idea of the value for moving forward, the firm should plan next to move strategically to the appropriate level. With at least an outline of the desired future state, the firm can establish a council or steering committee to construct the framework necessary for getting management endorsement and support for moving to the new state. That means an advanced supply chain model should be constructed, one that is enabled by collaboration and technology. Two caveats are in order. It becomes very important to make sure you know how to measure your progress so you can verify the value of the effort. It is also important to get some willing allies to help you build the model.

With a Fix on the Intended Level, Build the Model

With every industry believing it is unique, it is very hard to foster generic models for advanced supply chain efforts. The nucleus model presented earlier works as well as any in the beginning, but the most advanced firms have designed their own models. Exhibit 14.2 is a model for a high-technology components firm. From the simplified depiction of the flows between partners and internal operations, a company can build the necessary model

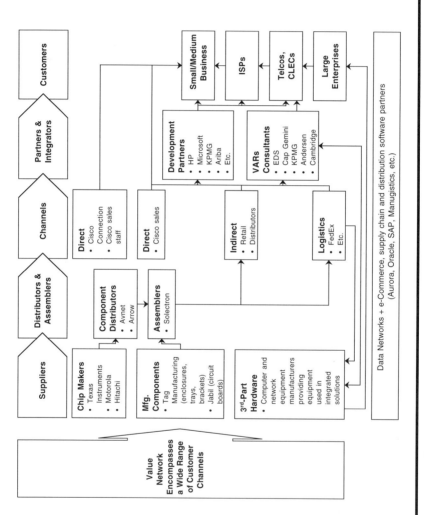

Exhibit 14.2 Supply chain process map.

with enough detail to begin process improvement. For all functions and business units, such models typically define what takes place in terms of product, information, and financial flow across what the company decides is the end-to-end total processing for that firm's business. The level of detail in the description of the model need not be deep in the beginning and can become more finite as progress is made.

For a steel manufacturer, that model could begin with iron ore, coal, limestone, manganese, and other minerals and proceed through blast furnace, steel making, casting, rolling, and delivery of coils to a fabricator. It could also extend to an OEM, like an automobile manufacturer. For a car maker, the model could go from suppliers of parts and subassembly manufacturers to assembly plants to a variety of distribution channels to dealer inventory and showrooms. The decision must be made based on the extent of coverage so the model brings attention to the resources that must be brought to bear on improvements in a prioritized manner. While we prefer to go as far downstream as possible, toward the end consumer, and also consider any returns that might occur, some firms do not need to go that far, at least in the early stages of their supply chain effort.

A consumer products firm is more advised to go all the way to the consumer than a chemical producer. The former type of firm builds a model that goes from manufacture through distribution, retail stores, and on to the consumer. That is a more elaborate model, but one that is crucial to getting the needs identified and satisfied for a much more fickle end constituency. A service organization can also construct such a model, perhaps working backward — from the intended services and level of satisfaction desired by targeted customer groups to what supply chain processing should take place.

With the help of willing and qualified partners, the firm then sets about to enhance the model, which should reflect how to attain above-industry supply chain capabilities. That requires working with the best-qualified suppliers, distributors, and key customers to define a state-of-the-art model that is as close to optimized operating conditions as possible and distinguishes itself in the eyes of the targeted customers/consumers.

Start on the Supply Side

Enhancing the model can start anywhere, but experience schools it is best to use the customers' need and satisfaction data as a guide and begin work on the supply side. The thinking is that the house should be in order before making promises that might not be kept. Using the help of the best suppliers, a template that reflects the current best practices is built and contains two features — the automation of routine process steps through e-procurement techniques, and superior supplier relationship management that benefits both parties. Recommendations on each aspect have been discussed in Chapter 11. The first feature will save time and effort from routine jobs and generally saves transaction costs and some improvement to indirect buying. The freed time can then be applied to more strategic efforts. The second feature is of greater importance and requires the partners to consider how the model can be enhanced for mutual benefit.

An immediate requirement will appear. Because nucleus firms will drive most efforts, and suppliers will find themselves aligned in several networks, it is crucial to spell out the parameters being involved in the collaboration. A basic ground rule applies: work with suppliers who will care as much about satisfying the business customer or consumer as you will, but give them an opportunity to build some strength for their firm. That means allow them some latitude to use the enhanced features of the collaboration as part of their business model, keeping truly proprietary features as an exclusive part of the arrangement.

In the most successful advanced efforts, buyers and suppliers come to an early agreement on what is exclusive and what can be used elsewhere. With that understanding, they progress to considering how to share resources and responsibility for improving the portion of the model where they have impact. That translates to providing qualified talent to think through how process steps can be improved so both parties benefit and something good happens for the intended customer. Eventually all of the key components of supply chain — forecasting, planning, developing new products, sourcing effectively, managing orders, storing products, and shipping products — will come under consideration. Use of resources will expand beyond the buying and selling personnel and include the desired vertical interface between multiple functions.

Inventory management, for example, becomes a joint effort. The actual stores of product are not something to push upstream in the supply chain, but become network assets. Joint initiatives are created as information is shared across an online network, so excess stocks are not necessary and the right goods are at the point of need. Business customers feed other data upstream on daily and weekly transactions. Now suppliers can match their capacities and stocks with what is being produced, so a final match is made with what is actually being demanded.

Firms collaborate with trusted suppliers to identify trends and requirements to make above-average consumer responses. The constituent having the greatest competency eventually handles each process step in the model so the total effort moves toward optimization. Under these circumstances there can be no tolerance of ineffective processing. The attention has to be brought to bear on how best to handle the end-to-end work with parties that come to rely on each other in ways not found in normal business processing.

Move Collaboration to the Delivery Side

With the help of the same qualified and willing suppliers, attention moves next to the delivery side. That includes distributors or any business partners that stand between the nucleus firm and the consumer. This determination should have been made during the construction phase of building the model. To maximize benefits, for example, the logistics function must be linked solidly to the processes that derive from the sourcing function and then extended forward to the parties responsible for getting the goods to the intended consumers. That means the best overall delivery system is established and the underlying premise of supply chain is met — the effort results in getting the right supplies and products to the point of need at the right time with the right quantity and quality all the time. It also means some effort has been made to optimize the shipment and storage costs for both buyers and sellers as the goods move through each link on the way to the customer.

The buyers and sellers, having completed their part of the model, turn to downstream planning and distribution. Issues considered at this point are order management, inventory management, fulfillment, warehousing, shipping, and financial flows. Working together, they set about to automate order processing, integrate planning systems, and build visibility into product flow and inventories. Again, recommendations on how to enhance these aspects have been presented in previous chapters. Partners work to minimize any buffer stocks, while keeping enough product flowing to the intended consumers. They use customer-oriented metrics to drive distribution and fulfillment. And they make sure everyone gets paid in a timely manner.

A word of advice is appropriate here. While much has been said about shorter cycle times, the key in building the network model is to have the appropriate cycles and then meet them religiously. Some situations are very well adapted to 5- or 7-day deliveries, for example, while others thrive on the shortest possible time frames, often down to hours. To shorten the first circumstance to hours may only increase buffer stocks rather than decrease them. The best scenario is to determine what fits the network needs and then keep to that time frame without fail. Advanced partners collaborate with each other to identify the optimum cycle time for the total network — from point of incoming materials to the actual sale and return, where appropriate.

To enhance the model being built, the partners attain access to downstream demand information as soon as possible, so supplies are matched with consumption. This condition requires network partners to share data on what is being pulled from the system all the way to the front end of the supply chain. The actual need for inventory is then identified and kept at levels that meet needs. With online visibility of what is in the chain, options are possible, including product diversions in time of special need.

This same capability allows the model designers to reconsider the warehouse or distribution center needs. Again, advanced partners look at dynamic flow characteristics to determine what products even need to go into storage and which can go direct to the next constituent in the supply chain. Viewing the flows online also enables the partners to adjust the flows, taking advantage of what was put into storage and blending it with what is in transit so customers are satisfied and as little inventory as possible sits around to gather dust.

Build Logistics Strength

As the model building progresses, the logistics function must take a central role in the effort, determining how to move product in the most effective manner. Using virtual logistics systems discussed previously, they build a depiction of the product flow and the inventories that reside between each process step. Collaborative logistics becomes an integral part of any successful supply chain network model.

Coordinating logistics across an end-to-end network has become one of the ultimate challenges to logistics professionals. Aligning activities and bringing them under some form of network control works well in theory, but applications are showing it is the real test of managing the distribution processes, particularly as the number of constituents increases. Time, resources, commitment, and patience are the operating ingredients. For these reasons, the partners involved must articulate the strategy in terms understood by all parties, determine which processes need to be enhanced to support the strategy, and how the operating model will create value for customers and those in the network.

We are considering here the heart of the supply chain strategy. If a nucleus firm is totally dedicated to being the low-cost producer, or focusing on operational excellence, then logistics will concentrate on costs, not time for delivery. On the other hand, if the firm is committed to customer intimacy, the partners could well agree that somewhat higher costs, by sending special trucks for next-day delivery or placing inventories closer to the user, are appropriate. The techniques must be matched with the inherent supply chain strategy. The result is a form of collaborative logistics that guides the storage and distribution of products to the consumer.

Collaborative logistics can be considered as the process of utilizing supply chain assets in the most effective manner for all parties. In that sense, these assets can be used effectively across a broad spectrum of entities, some directly involved in the network and some outside, so total utilization increases while providing extra features for network users. Several options appear as firms consider collaborative logistics. According to Karl Manrodt at Georgia Southern University, and Mike Fitzgerald, CEO of Elogex, Inc., a number of successful examples exist, including:

- Multienterprise continuous moves, or asset collaboration — achieving a lower contract transportation rate by linking together two or more product moves across an extended enterprise to eliminate non-paid miles for the carrier. Such a move increases asset utilization and lowers transportation costs.
- Collaborative less-than-truckload (LTL) consolidations — providing the visibility between network members to create truckloads from LTL shipments to enhance cost savings while sustaining shipment control.
- Warehouse collaboration — sharing excess warehousing capacity between network partners. Such a situation allows the partners to optimize warehouse utilization while reducing the premiums a single partner might pay for short-term storage.
- International container collaboration — providing drayage carriers with online visibility over containers moving to and from the port site. This condition enables the drayage carrier to optimize asset utilization and reduce operating costs, while shippers receive lower rates (Manrodt, 2001, p. 70).

There are other techniques available and more are being discovered as the logistics effort matures and blends with supply chain practices to find optimum conditions. The key is to work collaboratively with qualified partners and create visibility across the extended network. This visibility will provide the insight and information necessary to make the collaboration effective.

As one example, consider the alliance formed between network equipment maker Lucent Technologies, Inc. and Miami-based Ryder System, Inc. These partners are working with a few Lucent suppliers as part of a major initiative to improve visibility in the supply chain. According to Jim Schoessling, Lucent's senior manager for supply chain networks, the effort has resulted in "some unexpected benefits." This nucleus firm uses a system dubbed *Trade Stream* (software provided by Optum Inc.) to provide the suppliers with real-time information on demand and delivery dates. Suppliers using the system "can integrate this demand information with their production management systems to determine when they should begin producing components so they can deliver an order on time, without having to build inventory in advance and warehouse it until Lucent orders it" (Konicki, 2001, p. 49).

Ryder leads the logistics effort for the nucleus firm by applying Trade Stream to make certain each supplier can ship on schedule. If there is a problem, or if there is some kind of unexpected delay in a shipment, Ryder has the responsibility to redeploy material from other shipments or to find a different source so the material will be delivered on time. Complete orders are compiled at a Ryder facility for shipment to the appropriate Lucent plant or directly to a Lucent customer. Schoessling says Lucent should "realize significant savings when the software is deployed to nearly 100 key suppliers by year's end" (Konicki, 2001, p. 49).

As the logistics function is harmonized with the overall supply chain effort, the types of experiences cited become more common. As the combined effort moves to advanced levels of interaction, the necessity to enhance the ever more complex relationships through Internet technology becomes the real differentiator for success. Partners in the extended enterprise network simply have no choice but to refine their emerging model by introducing the technology that will define future state conditions.

Introduce the Necessary Technology through Collaboration

At the heart of the process of collaborating is the principle that all of the linked constituents in the extended network will benefit from joint solutions that meet or exceed the needs of the final consumer. Software suppliers are rushing in to help with those solutions, and an enormous new market is appearing before our eyes. AMR Research estimates the collaboration software market will grow from $1.1 billion in 2000 to a staggering $8.7 billion by 2004 (O'Marah, 2001, p. 23).

While I endorse the application of software once the collaborators have defined how it will enhance network capability, most of the applications have been for enhancing internal excellence, or Level 2 performance. The time has come to move the application to a higher level and focus on overall network performance. But software is only one aspect of the solutions being sought by inter-enterprise partners. And these partners must first decide on how to put the technology to good use. When done well, the results can be staggering. The computer industry, for example, a frontrunner in the supply chain evolution, has used collaborative manufacturing techniques enhanced with technology to achieve productivity growth averaging 4.6% per year for 15 years.

Let us consider one element of supply chain collaboration that can lead to such improvements. One of the biggest unsolved problems in supply chain processing is due to the incessant changes made to what look like optimized plans and schedules. These interruptions and subsequent changes can cost a firm as much as 10% of total revenue, according to supply chain strategist Kevin O'Marah at AMR Research. According to O'Marah, "Some of this waste is the result of poor forecasting. But a substantial portion represents buyers' hedging their needs and pushing uncertainty back to suppliers. Collaboration value here is a product of full disclosure." He estimates that in the U.S. manufacturing economy alone, "the incremental operating margin achievable through collaboration could be as high as $390 billion per year" (O'Marah, 2001, p. 24).

With such an opportunity as a backdrop, we submit any successful collaboration strategy must include a viable technology platform. Central to that platform must be software selected by network partners that provides the online information necessary to improve decision making at each point of transfer between partners. The portion of the model being developed between network partners must depict where the technology will be applied, how it will enhance performance, who has responsibility for control, and how the final consumer benefits. How the subsequent decision making will be better and how it will be explained across the network should also be a part of the systems being designed into the model. This requirement means the model designers will explain who has access to the information and how it will be applied to network decisions.

As an example, consider one of the most highly visible applications struggling to attain acceptance and proven value — collaborative planning, forecasting, and replenishment (CPFR). This application could be the mother lode for supply chain advocates if technology can be satisfactorily applied across networks. CPFR is characterized by the need for highly structured information sharing and use of that information in better decision making. Supply capabilities, demand data, inventory status, logistics information, and actual consumption are combined in advanced models. With better information, the involved parties make specific commitments across what becomes the visible structured workplace. Product flow is tracked online and the right inventories move to the correct point of need in the right quantities. It is a wonderful model. It needs coordination of the enabling technology with network consensus.

Partnering Is the Key in the New Economy

Collaborative business networks enhanced by technology deliver superior value to customers, consumers, partners, and shareholders. Those who build these networks will outperform other networks and dominate their industries. To succeed, companies across the extended enterprise must consider and understand what each does best and how the total effort can be shared to come as close as possible to optimum conditions. Together they must introduce information standards that will differentiate their network from competitors and build a platform together that benefits all constituents.

In the networked economy, relationships are crucial but they must be strategic. The strategies can no longer be a purely internal matter. It must be something that involves trusted partners with similar technological capability. With the basis of the new competition being oriented around knowledge, only those networks with partners sharing the latest and most valuable information through a real-time platform will succeed.

Unfortunately, relationships like those being advocated are seldom simple to form or sustain. Each partner naturally has self-interest in survival and prosperity. But all firms today are part of an extended enterprise network, and there is a necessity to rely on at least a few trusted allies so you can build a superior e-business model. Central to that model will be the real source of competitive advantage — knowledge applied across the end-to-end supply chain.

Now is the time to build your knowledge assets with the help of these external resources. Frameworks that clarify what sort of relationships meet the needs of the end consumers need to be constructed. These frameworks must be coordinated between network partners and enhanced with technology. This form of partnering requires attainment and use of the skills described throughout this text. Moreover, this partnering is the only way to find the roadway to future success.

Summary

There can be no doubt that use of the Internet will ultimately transform whole industries, redefining the way business is conducted, and introducing the mechanisms that will allow firms in an extended enterprise to collaborate and compete more effectively. Using new patterns of interaction, which have been enhanced by cyber technology, network partners will enrich the value of traditional interactions as they focus on the real needs of the customers and consumers they are satisfying. This enrichment will be based on online communications and sharing that result from the collaboration among trusted allies.

The steps forward are fairly clear. They require selecting a few willing and capable partners to define a business model that meets or exceeds the needs of the targeted consumers. Together the partners consider every point of interaction and bring the right technology applications to enhancing those interactions for the mutual benefit of all partners and their ultimate consumers. They begin by analyzing what the customer and consumers want, then identifying the resources of knowledge, and working with partners to build the superior model and means of making the best any competitive network.

References

Agrawal, Mani and Pak, Minsok, "Getting Smart about Supply Chain Management," *The McKinsey Quarterly*, 2001, No. 2: On-Line Tactics, pp. 22–25.
Ante, Spencer, "Simultaneous Software," *Business Week*, August 27, 2001, pp. 146–147.
Berry, Leonard, "The Old Pillars of New Retailing," *Harvard Business Review*, April 2001, pp. 131–137.
Carman, Rick, "Key Performance Indicators: Putting the Customer First," *Supply Chain Management Review*, November/December 2000, pp. 90–95.
Carter, Larry, "Cisco's Virtual Close," *Harvard Business Review*, April 2001, pp. 22–23.
Champy, James, "Why Partnerships Fail," *Computer World*, July 30, 2001, pp. 22–25.
Chen, Anne, "ERP: It's Alive," *PC Week*, May 6, 2001.
Chopra, Sunil, Dougan, Darren, and Taylor, Gareth, "B2B e-Commerce Opportunities," *Supply Chain Management Review*, May/June 2001, pp. 50–57.
Colkin, Eileen, "Owens Corning Gets Personal with Business Customers," *Information Week*, July 30, 2001, p. 26.
Colkin, Eileen, "Personalization Tools Dig Deeper," *Information Week*, August 27, 2001, pp. 49–50.
Copeland, Ron, "Heeding Customer Requests — Voice Recognition," *Information Week*, July 9, 2001, pp. 45–46.
Crawford, Fred and Mathews, Ryan, *The Myth of Excellence*, Crown Business, New York, 2001.
Degraeve, Zeger and Roodhooft, Filip, "A Smarter Way to Buy," *Harvard Business Review*, June 2001, pp. 22–23.
Durtsche, David, Keebler, James, Ledyard, Michael, and Manrodt, Karl, *Keeping Score: Measuring the Business Value of Logistics in Supply Chain*, Council of Logistics Management, Oak Brook, IL, 1999.
Engardio, Pete, "Smart Globalization," *Business Week*, August 27, 2001, pp. 132–138.
Ferrari, Robert, "Frontline Supply Chain Managers Struggle with E-Business," Richmond Events and AMR Research White Paper, AMR Research, Boston, MA, September 2000.
Foster, Thomas, "It's Back to Basics in the B2B World," *Supply Chain e-Business*, February 2001, p. 8.

Fryer, Bronwyn, "High Tech the Old-Fashioned Way," An Interview with Tom Siebel, *Harvard Business Review*, March 2001, pp. 119–125.

Gartner, Inc., *CRM Solutions Outlook: A Look Across Vertical Markets*, June 20, 2001, Gartner, Inc., Stamford, CT.

Ghemawat, Pankaj, "Distance Still Matters," *Harvard Business Review*, September 2001, pp. 137–147.

Gooley, Toby, "Enter the New Dragon," *Supply Chain Management Review*, November/December 2000, pp. 17–18.

Gooley, Toby, "Crafting a Latin American Strategy," *Supply Chain Management Review*, May/June 2001, pp. 17–18.

Gould, Stephen, "Sourcing Successfully in China," *Supply Chain Management Review*, July/August 2001, pp. 44–54.

Haddad, Charles, "Office Depot's E-Diva," *Business Week*, August 6, 2001, pp. EB22–24.

Hamm, Steve, Welch, David, Zellner, Wendy, and Engardio, Peter, "E-Biz: Down But Hardly Out," *Business Week*, March 26, 2001, pp. 126–130.

Herrell, Elizabeth, "Managing Customers in the New Economy," *Conspectus*, May 2001, pp. 16–17.

Huen, Christopher, "Delivery, Anyone?" *Information Week*, July 16, 2001, pp. 22–24.

Kaiser, Rob, "Human Touch Selling Online," *Chicago Tribune*, Business Section, September 10, 2001, pp. 1–4.

Kambil, Ajit and Sparks, Scott, "Seizing the Value of e-Procurement Auctions," *Supply Chain e-Business*, February 2001, pp. 53–54.

Karpinski, Richard, "FedEx's E-Biz Philosophy," www.internetweek.com, July 11, 2001, pp. 1–4.

Karpinski, Richard, "J.B. Hunt's EDI Swap-Out," www.internetweek.com, August 15, 2001, pp. 1–3.

Karpinski, Richard, "Goodyear Explores E-Business," www.internetweek.com, August 9, 2001, pp. 1–3.

Karpinski, Richard, "The Limited's Big Picture," www.internetweek.com, August 23, 2001, pp. 1–4.

Keenan, Faith, "Opening the Spigot," *Business Week*, June 4, 2001, pp. EB17–20.

Kehoe, Louise, "Time for Chief Executives to Become e-Literate," *Financial Times*, January 30, 2001, pp. 1–3.

Kemp, Ted, "Amazon Strikes Electronics Partnership," www.internetweek.com, August 20, 2001a, pp. 1–2.

Kemp, Ted, "Departmental Agendas Diverge," www.internetweek.com, August 20, 2001b, pp. 1–5.

Koudsi, Suzanne, "Consultants: Who Are We," *Fortune*, September 3, 2001, p. 48.

Konicki, Steve, "Ryder's Movin' On," *Information Week*, June 25, 2001, pp. 39–50.

Lapide, Larry, "Buyers and Sellers Evolving into Collaborators," *Business Week*, March 26, 2001, Special Advertising Section.

Leon, Mark, "Expensive! Complicated! Takes on a Life of Its Own! But You Can't Escape from CRM," *Infoworld*, July 13, 2001, pp. 1–8.

Lewis, David, "Intranet Offer Helps Big Boy Restaurants Pick Pepsi," www.internetweek.com, August 13, 2001, pp. 1–2.

Ljungdahl, Lars, "What You Need to Know about the Internet-enabled Supply Chain," *Supply Chain Management Review*, November/December 2000, pp. 82–88.

Magidson, Jason and Brandyberry, Gregg, "Putting Customers in the 'Wish Mode,'" *Harvard Business Review*, September 2001, pp. 26–28.
Manrodt, Karl B. and Fitzgerald, Mike, "Seven Propositions for Successful Collaboration," *Supply Chain Management Review*, July/August 2001, pp. 66–72.
Maselli, Jennifer, "People Problems," *Information Week*, July 9, 2001, pp. 35–76.
McDougall, Paul, "Collaborative Business," *Information Week*, May 7, 2001, pp. 43–66.
McDuffie, John, West, Scott, Welsh, John, and Baker, Brent, "Logistics Transformed: The Military Enters a New Age," *Supply Chain Management Review*, May/June 2001, pp. 92–98.
McGee, Marianne Kolbasuk, "Supply Chain's Missing Link," *Information Week*, September 17, 2001, pp. 39–41.
Mentzer, John T., Roggin, James H., and Golicic, Susan L., "Collaboration: The Enablers, Impediments, and Benefits," *Supply Chain Management Review*, September/October 2000.
Moozakis, Chuck, "GE Scales Back," *Internet Week Online*, May 10, 2001, pp. 1–4.
Neuborne, Ellen, "The Box that Rocks," *Business Week*, June 4, 2001, p. EB6.
O'Marah, Kevin, "A Reality Check on the Collaboration Dreams," *Supply Chain Management Review*, May/June 2001, pp. 23–26.
Pazmany, Peter, "Sun's Virtual Network," *Supply Chain Management Review*, November/December 2000, pp. 56–61.
Pearlson, Keri, "Doing Business @ the Knowledge Level: Instant Involvement," *CSC Foundation Research Journal*, March 2001, pp. 74–81.
Poirier, Charles and Bauer, Michael, "Toward Full Network Connectivity," *Supply Chain Management Review*, March/April 2001, pp. 84–90.
Porter, Michael, "Strategy and the Internet," *Harvard Business Review*, March 2001, pp. 63–78.
Power, Denise, "Crossing the Channels," *Executive Technology*, July 2001, pp. 14–20.
Power, Denise, "Saks White Paper Asks: Whither CRM?" *Executive Technology*, July 2001, p. 11.
Quinn, Frank, "Collaboration: More than Just Technology," *The Aspect Project*, Accenture Publication, New York, 2000, pp. 222–223.
Radjou, Navi, "Deconstruction of the Supply Chain," *Supply Chain Management Review*, November/December 2000, pp. 30–38.
Schultz, Keith, "SCM Turned Inside Out," www.internetweek.com, June 25, 2001, pp. 1–9.
Sedlack, Patrick, "The Second Wave of e-Fulfillment," *Supply Chain Management Review*, May/June 2001, pp. 82–88.
Seldon, Larry and Colvin, Geoffrey, "Will Your E-Business Leave You Quick or Dead?" *Fortune*, May 28, 2001, pp. 112–124.
Seybold, Patricia, "Get Inside the Lives of Your Customers," *Harvard Business Review*, May 2001, pp. 80–89.
Smith, Tim, "Don't Blame the Internet for Retailers' Woes," www.internetweek.com, August 1, 2001, p. 1–2.
Smock, Doug, "Get Set to Board the Productivity Express," *Supply Chain Management Review*, May/June 2001, pp. 113–114.
Stank, Theodore P., Frankel, Robert, Frayer, David, Goldsby, Thomas, Keller, Scott, and Whipple, Judith, "Tales from the Trenches," *Supply Chain Management Review*, May/June 2001, pp. 62–69.

Stone, Floyd, "Assessing the Variables of Global Trade," *Supply Chain Management Review*, July/August 2001, pp. 9–10.

Stuart, John, "Long and Winding Road to Online Shopping Therapy," *Conspectus*, May 2001, pp. 46–47.

Vijayan, Jaikumar, "Procurement Network, Harnessing Buying Power," *Computer World*, July 4, 2001, pp. 18–22.

Walton, Michael, "Lacking an e Drive," *Conspectus*, May 2001, pp. 24–25.

Watson, James K., "The Value of Collaboration," www.informationweek.com, listening post, July 30, 2001, pp. 1–2.

Whiting, Rick, "Carrier Fans Sales with Data Mining," *Information Week*, August 6, 2001, p. 55.

Wilder, Clinton and Soat, John, "The Trust Imperative," *Information Week*, July 30, 2001, pp. 34–42.

Wilson, Tim, "Global E-Biz Mishmash," www.internetweek.com, August 27, 2001, pp. 1–3.

Index

A

Alliances
    cooperative behavior in, 198
    examples of, 192–193, 224
    failure of, 192
    formation of, 191–194
    globalization, 126
    need for, 192
Amazon.com, 60–61

B

Boeing, 174–175
Briggs & Stratton, 113
Built-to-order manufacturing, 57
Business needs, 72–74
Business partners, collaboration among
    customer satisfaction, 60–61
    Internet ventures, 99–100
    technology, 61–62
Business plan, 18
Business-to-business e-commerce
    collaborations, 156
    description of, 101–102, 117
    expenditures in, 110
Buyer
    collaboration participation by, 113–117, 200–203
    seller and, relationships between, 117, 155, 220
Buying experience for customers, 83–84

C

Calibration of supply chain, 210–216
Carrier Corp., 58
Case studies
    Cisco Systems, 22–25, 103
    collaboration, 52–53
    consumer products company, 36–37
    customer focus, 90–92
    customer relationship management, 74–77
    Gallery Furniture, 90–92
    globalization, 129–131, 140–146
    Hitachi Europe Ltd., 140–146
    Intel, 105–107
    Internet, 105–107
    metrics, 36–37
    Moen, Inc., 52–53
    Procter & Gamble, 61–62, 74–77, 114, 198
    sourcing, 166–170
CDW Computer Center Inc., 86–87
CFO, 20
Chief purchasing officer, 166–167
CIO, 20–21
Cisco Systems, 22–25, 103
Collaboration
    activities necessary for, 113–114
    assessment of readiness for, 44, 201–203
    benefits of, 195
    business partners, 60–61
    buyer's participation in, 113–117, 200–203

case studies of, 52–53, 112, 122–123, 196–197, 199–200, 203–207
constituent benefits, 49
cooperation established through, 42–43
cooperative behavior in, 198
criteria for success, 45, 47
cross-functional, 70
definition of, 102
delivery side, 221–222
design partners, 50, 59
development of, 112
enablers for, 195–197
end result of, 43
expectations for, 198
external advice, 42
impediments to, 51–52
importance of, 39–40, 109, 174
information sharing, 47
international container, 223
Internet opportunities for, 43
limitations of, 174–177
logistics, 223
need for, 41–42
nucleus firm in
   buy side, 49–50
   description of, 49
   product side, 50–51
   sell side, 51
   sourcing, 157
participants in, 113–117
partnering diagnostic laboratory, 118–121
personnel necessary for, 52
preparedness chart for, 201–202
seller's participation in, 113–117
software, 225
summary overview of, 53–54
technology for
   case study use of, 62–63
   description of, 61–62
   introduction of, 224–226
trust in, 195, 197–200
value perspective emerging from, 47–49
warehouse, 223
Collaborative commerce, 43
Collaborative planning, forecasting, and replenishment, 225–226
Consolidated sourcing, 164
Consumers, *see* Customers
Continuous improvement, 173
COO, 20
Cooperation, 42–43
Cost/productivity improvement, 19
Cost savings tracker, 168
Culture
   description of, 173
   globalization considerations, 132
   hidden values for, 183–187
   inhibitors, 177–178
   leadership considerations, 180–181
   supply chain concepts effect on, 177
   supply chain model, 182–183
   testing of potential improvements, 187–189
   training steps, 179–180
Customer(s)
   business interfaces with, 58–60
   buying experience for, 83–84
   case study of, 90–92
   changing patterns of, 57
   collaboration benefits for, 49
   customization appeal for, 57, 90
   data regarding, 81–82
   design involvement by, 86
   desires of, 56, 83–85
   e-fulfillment of orders, 89–90
   focus on, 4, 56–58, 83, 85–87
   full service provided to, 85
   identifying of, 209
   Internet purchasing by, 83–85
   inventories matching demands of, 59–60
   knowledge of, 81–82
   loyalty of, 57
   marketing efforts directed at, 82
   metrics designed for, 27, 30–31, 34–35
   multichannel nature of, 87–88
   needs of, 79–81, 209, 219
   purchasing preferences, 56, 83–85
   returning of products by, 84
   satisfaction of, 28, 84
   strategies focusing on, 85–87
   supplier satisfaction of, 61
   supply chain responsiveness to, 60
Customer care, 214–215
Customer relationship management
   benefits of, 71
   business needs focus of, 72–74
   case study of, 74–77

criteria for success, 72–74
customer-centric strategy, 73–74
definition of, 68, 71
description of, 65
failure of, 66–67, 77
history of, 65
implementation of, 71
inherent aspects of, 68
issues regarding, 74
marketing integration, 70
objectives of, 68, 73
prevalence of, 70
strategies for, 67–70
summary overview of, 77
technology for, 69–71
Customization, 57, 90

D

Data mining, 58
Data sharing, 32
Delivery system, 221–222
Digital technology, 186
Diminishing returns, 55, 150
Direct spending efficiencies, 16

E

E-business opportunities, 184–187
E-commerce
    Britain participation in, 176
    business-to-business
        collaborations, 156
        description of, 101–102, 117
        expenditures in, 110
    development of, 174
    evaluation of, 186
Economic valued added, 185–186
E-fulfillment, 89–90
Electronic data information, 161, 163
Employees
    talented types of, 181–182
    vision shared with, 21
Engineering, 212–213

Enterprise resource planning, 71, 114, 178, 185
E-procurement, 157–162, 220
E-retailing, 88
Ethics, 197–198
Evolution, of supply chain
    description of, 6–7
    illustration of, 6
    levels in, 2–5
    results of, 7
Extended enterprise
    case study of, 122–123
    collaborations, 116–117, 122–123
    description of, 104–105
    designing of, 182
    information sharing, 110–113
    maximizing of, 104
    steps involved in, 115–116
    training efforts, 180
External advice, 42
Extranet, 113, 196

F

Fulfillment
    e-based, 89
    traditional, 89

G

General Electric, 160
General Mills, 156
General Motors, 105
Globalization
    buying power considerations, 130
    case study of, 129–131, 140–146
    company alliances for, 126
    cultural considerations, 132
    delivery considerations, 139
    demographic considerations, 127–129
    difficulties associated with, 125–129
    distance considerations, 131–133
    distribution centers, 135–136
    infrastructure considerations, 132–133
    manufacturing competencies, 138
    network construction, 137

process review, 137–140
reasons for, 127–128
requirements for, 126
shipping, 136
sourcing, 169–170
standards that affect, 133–134
summary overview of, 146–147
supply chain demands, 125, 128–129, 138
technology support, 139–140
transportation services, 135–136
trends in, 126
virtual organization, 139
Web technology centralization, 134
World Trade Organization, 131
Global purchasing process, 170–171
Global resource planning, 200
Goodyear, 196

H

Human resources, 215

I

IKEA, 85–86
Improvements, 3, 15
Indirect material sourcing, 159
Information sharing
    Internet, 110–113
    trust and, 197
Information technology, 216
Infrastructure, 203
In-store kiosks, 88
Inter-enterprise synchronization, 5, 45, 117, 161
Internal constituents
    collaboration
        assessments of readiness, 44, 201–203
        benefits of, 49
        coordination of, 32
International container collaboration, 223
Internet
    benefits of, 98, 183
    business collaborations, 99–100
    business manager's views of, 175

business processes enhanced by, 103–104
business-to-business e-commerce, 101–102
business value of, 99
capitalization of, 96–97
case study of, 105–107
collaborations using, 43, 52–53
customer purchases on, 58, 83–86
design collaborations using, 52–53
failures in, 96–97
framework for working with, 101
growth of, 125
history of, 94–96
information sharing, 110–113
knowledge about, 98
limitations of, 174–177
market efficiencies, 102
phone sales combined with, 87
supply chain effects, 102, 104–105, 176
trends in, 94–96
understanding of, 176
value chain benefits of, 98
value of, 103, 110
Intra-enterprise process, 34
Inventory
    customer-specific fulfillment of, 59–60
    management of, 220
    reduction of, 15–16
Ito-Yokado, 81

J

Joint assets, 151
Just-in-time efforts, 115

L

Leaders
    convincing of, 180–181
    external collaboration resisted by, 41–42
    organizational scrutiny of, 181
    reasons for adopting supply chain, 11–14
    vision aligned with, 20–21
Less-than-truckload consolidations, 223
Limited, The, 88, 199
Limited Technology Services, 199

Index ■ 237

Logistics
  building of, 222–224
  characteristics of, 213
  collaborative, 223
  coordinating of, 222
  metrics for, 34
  reduced costs for, 16

M

Manufacturers, 45
Manufacturing
  built-to-order attention in, 57
  calibration of, 212–213
Margin increases, 17
Marketing, 212
Metrics
  action study of, 36–37
  changing of
    process of, 31–34
    reasons for, 34–35
  customer relations, 27, 30–31, 34–35
  description of, 27
  internal excellence focus of, 27
  net sales, 29
  performance, 28–31
  procurement-related, 32–33
  pull system, 29–30
  push system, 29
  summary overview of, 37
  traditional, 27–31
Model for supply chain, 182–183, 217–219
Multienterprise continuous moves, 223

N

Nestle Corp., 61
Net profits, 55
Net sales, 29
Network connectivity, 5, 162–163
Networked supply chain, 43
Nucleus firms, in collaborative relationships
  buy side, 49–50
  description of, 49, 114
  product side, 50–51
  sell side, 51
  sourcing, 157, 164

O

Office Depot, 62–63
Online marketplaces, 165
Opportunities
  added-value, 40
  business-related, 40–42
  savings-related, see Savings opportunities
Orders
  customization of, 90
  e-fulfillment of, 89–90
  tracking of, 90
Owens & Minor, Inc., 69

P

Partnering diagnostic laboratory, 117–121
Partnerships
  case study of, 203–207
  collaborative
    customer satisfaction, 60–61
    Internet ventures, 99–100
    technology, 61–62
  failure of, 193
  framework for, 226
  importance of, 226–227
  partnering assessments, 203
  strategic, 226
Penske Logistics, 113
Performance metrics, 28–31, 185–186
Personalization, 88
Planning, 212–213
Processing time, 33
Procter & Gamble, 61–62, 74–77, 114, 198
Procurement
  description of, 32
  e-procurement, 157–160, 220
  internal optimization of, 211
  metrics for, 32–33
Progression
  bottom line focus effects on, 55
  steps involved in, 8–9
Pull system, 29
Purchasing
  customer preferences, 56, 83–85
  description of, 211
  Internet-based, 83–85
Push system, 29

R

Reciprocal marketing, 159
Recreation Equipment, Inc., 84
Research and development, 17
Returning of products, 84
Return on investment, 186
Return on net assets employed, 186
Revenue
    business partner collaborations for, 60–61
    e-procurement for, 161
    focus on, 55
    technology benefits for, 71
Reverse logistics, 2
Road Runner Sports, 82

S

Sales
    calibration of, 212
    savings opportunities for, 17
Samsonite, 82
Savings opportunities
    description of, 14–15, 40
    direct spending efficiencies, 16
    indirect spending, 16
    inventory reduction, 15–16
    logistic costs reduced, 16
    margin increases, 17
    new business, 17
    research and development, 17
    sales-related, 17
    vision supported by, 18–20
Scheduling, 212–213
Seller-buyer relationships, 117, 155, 220
Shared services, 4
Sourcing
    advanced supply chain planning, 211
    alignment with other companies, 164
    benefits of, 160
    case study of, 154–155, 166–170
    collaborative relationships, 153
    consolidated, 164
    cost reduction through, 150
    description of, 149
    electronic buying, 162–166
    e-procurement, 157–162
    framework for, 162
    global, 169–170
    hosting services, 165
    improvements in, 165
    indirect material, 159
    internal optimization of, 211
    network connectivity, 162–163
    nucleus firms in, 157, 164
    online marketplaces, 165
    participation costs, 165
    progression of, 152
    savings associated with, 149
    staffing reductions, 152–153
    strategic considerations, 151
    suppliers as partners, 154–157
Strategic partnerships, 226
Supplier(s)
    business manager's expectations, 175
    changes in, 45
    collaboration involvement, 200–203
    inventory and sales data shared with, 61
    joint ventures, 159
    partner status of, 154–157
    precise delivery by, 175
    technical skills, 178
Supplier base, 33
Supplier use index, 33
Supply chain
    advanced management of, 175
    awareness of, 177
    business opportunity of, 40–42
    calibration of, 210–216
    complicated nature of, 44–47
    costs of, 111, 152
    description of, 1
    effectiveness of, 7–8
    evolution
        levels of, 2–7
        need for, 39
    improvements offered by, 3, 15
    Internet effects, 102, 104–105, 176
    map for, 218
    model of, 182–183, 217–219
    networks, 111–112, 116
    optimization of, 1–2, 8, 211–216
    participants in, 45–46
    potential of, 14–18
    premise of, 29
    profit-making opportunities, 13

progression in, 8–9, 209–210
reasons for adopting, 11–14
responsiveness of, 60
schematic diagram of, 46
scope of, 2
steps involved in, 21–22
success of, 30
summary overview of, 9–10
technology for, 61–62
traditional view of, 40–41
training regarding, 179–180

T

Technology
    collaborative
        case study use of, 62–63
        description of, 61–62
        introducing of, 224–226
    consolidation of, 200
    customer relationship management, 69–71
    importance of, 191
    Internet, *see* Internet
    sales increases using, 66
Testing of potential improvements, 187–189
Trading exchanges, 163, 165
Trading partners, 191
Transactional relationships, 117
Transaction processing, 33
Transportation needs, 4
Trust
    case study of, 203–207
    in collaborations, 195, 197–200
    ethics and, 197–198
    external, 195
    formation of, 199
    information sharing, 197
    internal, 195
    need for, 195

V

Value-adding features, 116–117
Value chain
    collaborative partners effect on, 47–49
    cyber technology as enabler of, 5
    description of, 5
    Internet benefits for, 98
Value discovery, 202
Vendor relationships, 23–24
Virtual organization, 139
Vision
    creating of, 20
    key players aligned with, 20–21
    savings opportunities as providing support for, 18–20
    sharing with employees, 21
    sustaining of, 21–22

W

Wal-Mart, 74–77, 83, 102–103
Warehouse collaboration, 223
Wisdom, 69–70
World Trade Organization, 131